Methods in DNA Amplification

Methods in DNA Amplification

Edited by

Arndt Rolfs,
Ines Weber-Rolfs, and
Ulrich Finckh
Free University of Berlin
Berlin, Germany

Springer Science+Business Media, LLC

Library of Congress Cataloging-in-Publication Data

Methods in DNA amplification / [edited by] Arndt Rolfs, Ines Weber
-Rolfs, and Ulrich Finckh.
      p.   cm.
   "Proceedings of the Second International PCR Symposium on Usage of
PCR and Alternative Amplification Methods in Infectious and Genetic
Diseases, held February 26-27, 1993, in Berlin, Germany"--T.p.
verso.
   Includes bibliographical references and index.
   ISBN 978-1-4613-6078-0   ISBN 978-1-4615-2530-1 (eBook)
   DOI 10.1007/978-1-4615-2530-1
   1. Polymerase chain reaction--Congresses.   I. Rolfs, Arndt, 1959-
. II. Weber-Rolfs, Ines|. III. Finckh, Ulrich. IV. International
PCR Symposium on Usage of PCR and Alternative Amplification Methods
in Infectious and Genetic Diseases (2nd : 1993 : Berlin, Germany)
   [DNLM: 1. Polymerase Chain Reaction--methods--congresses.   2. DNA-
-genetics--congresses.   3. Hereditary Diseases--diagnosis-
-congresses.   4. Communicable Diseases--diagnosis--congresses.   QH
450.3 M592 1994]
QP606.D46M48   1994
574.87'3282--dc20
DNLM/DLC
for Library of Congress   94-43082
                                                           CIP

Proceedings of the Second International PCR Symposium on Usage of PCR and Alternative
Amplification Methods in Infectious and Genetic Diseases, held February 26–27, 1993,
in Berlin, Germany

ISBN 978-1-4613-6078-0

© 1994 Springer Springer Science+Business Media New York
Originally published by Plenum Press, New York in 1994
Softcover reprint of the hardcover 1st edition 1994

PREFACE

The polymerase chain reaction (PCR) - an in Vitro techniques for producing large amounts of a specific DNA fragment - has rapidly become established as one of the most important, impressive and fascinating methods of molecular biology as well as clinical diagnostics. In the seven years since' the technique was published, it has had a major impact on medical research. However, as there are still problems in instruments, standardized protocols for diagnostic applications and unsolved difficulties to avoid cross-contaminations on the one hand and on the other hand the even present question of how to interpret the biological value of a PCR-result, most clinicians prefer to further wait until these topics are clarified.

It is the aim of this book to give the reader lab-proven protocols from experienced scientists as well as a general introduction to alternative DNA-amplification procedures and their possible usage such as the NASBA or LCR.

This book is divided into four major parts to provide a theoretical (first and second section) and a practical framework for a better understanding of the new technology. In the first part we provide an up-to-date summary of basic problems in this rapidly evolving field. We demonstrate, for example how to use fixed tissue materials and how to quantify PCR products as well as how to prepare nucleic acids in a safe, convenient and proper way, or even how to sequence directly PCR products for the analysis of the DNA structure. The second part of this book is a short compilation of alternative methods for DNA amplification procedures which may be better than the PCR in some circumstances. The ligase-mediated detection techniques seem to be especially robust. In the third part detailed protocols and data are given to the broad field of virus detection. The early and safe diagnosis of viral infections is increasingly important in the work of the clinical laboratory, mainly as the number of immunocompromised patients rises. In this section we have tried to give a representative but by no means exhaustive compendium of techniques for the specific detection and characterization of viral pathogens. In most articles the problems of the biological significance of PCR-results are discussed very carefully. In the last section bacterial and fungal-pathogen data are given. In some circumstances, since these infections can be life-threatening in a very short period of time, rapid and accurate diagnosis - for example of mycobacterial infections - is of great importance in ensuring proper patient care.

The results of the investigations in the different and rapidly growing field of PCR technology give a lot of answers to actual questions. But with each new answer, much more new questions arise. This book, made to address such questions, is the product of a symposium dealing with up-to-date problems on techniques in the field of DNA amplification technologies, mainly the polymerase chain reaction but also alternative techniques. In February 1993, the

2nd International PCR Meeting took place in Berlin, Germany and the invited speakers were asked to submit a manuscript concerning their topics so that the book could be published with their recent scientific data. This book is a result of such an opportune undertaking. The editors hope that this volume will provide something of interest to everyone; to those new to the PCR yet willing to start with this technique, but also to those already well familiar with the PCR field.

The editors would like to thank Joanna Lawrence, Nicola Clark and Thomas Lewis Flood from Plenum Press for their permanent help, constant readiness to give information and infinite patience, Katrin Schmidt for preparing most parts of the manuscripts as well as Loraine Davis for some linguistic revisions. Last but not least, the editors thank all the authors for submitting their manuscript and, thus, contributing to the success of the book.

Berlin, May 1994   Arndt Rolfs
                                                                                             Ines Weber-Rolfs
                                                                                             Ulrich Finckh

CONTENTS

General Aspects of PCR

Alternative DNA-Amplification Methods

PCR and Virological Problems

PCR in the Field of Bacterial and Fungal Problems

General Aspects of PCR

THE POLYMERASE CHAIN REACTION AND FIXED TISSUES

John O'Leary

Nuffield Dept. of Pathology and Bacteriology, University of Oxford, UK

INTRODUCTION

The polymerase chain reaction (PCR) is a powerful tool for the retrospective analysis of fixed paraffin wax embedded material. This is the commonest source of tissue for the histopathologist. However, amplification failure is now observed in some centres using fixed tissues with PCR. Rates vary from 2% to 30% (An and Fleming 1992). Many of these failures are due to direct fixative interactions with nucleic acid templates. In this article, I will discuss the interaction of nucleic acids and fixatives and apply the findings to explain the results obtained with fixed tissues when using PCR.

Nucleic acid and fixative interactions - how they influence PCR

Nucleic acid and fixative interactions are varied. Many fixatives have been used for preservation of nucleic acids in tissue specimens, but relatively few with the exception of mercury and chromium salts are known to react with them chemically.

Fixative type and fixation time have long been known to influence the preservation of proteins for immunohistochemistry (Battifora et al 1986). It is now clear from our own work and similar observations (Greer et al 1991, Barton-Rogers et al 1990) that they also directly influence the results of PCR. Formal saline, 10% formalin, neutral buffered formaldehyde, Carnoy's and glutaraldehyde fixed tissues usually yield amplifiable DNA.

Formal saline, 10% formalin, neutral buffered formaldehyde and glutaraldehyde are members of the formaldehyde group of fixatives. These fixatives induce extensive cross-linking between nucleic acids and proteins. DNA and RNA in their resting native states do not normally react with these fixatives. However, if the fixative solution is heated to 45°C for DNA and 65°C for RNA, uncoiling of the helices occurs. This selective non-reactivity appears to be related to the structure of the nucleic acids, the hydrogen bonded structures of which are only broken at relatively high temperatures when purine and pyrimidine bases become available for reaction with the aldehyde fixative (McGhee JD et al 1975a and b, 1976a and b). These reactions may

Methods in DNA Amplification, Edited by
A. Rolfs *et al.*, Plenum Press, New York, 1994

be reversible or irreversible. Depending on the degree of cross-linkage then, it seems conceivable that erratic PCR results may be obtained from tissues fixed with one of the formaldehyde fixatives.

The alcohol fixatives (ethanol and methanol) are known to preserve DNA and histoneproteins. From our own observations, ethanol fixed tissue yields good results with proteinase K extraction, which is in accordance with similar findings by others (Greer et al 1991, Jackson et al 1990).

Bouin's (picric acid fixative) fixed tissue yields variable results. Picric acid derivatives are known to react with histoneprotein subfractions and basic proteins to form picrates. DNA bound histoneproteins which are intercalated with picric acid residues are then unavailable for reaction in the PCR, and thus amplification failure can occur.

Disappointing results are also observed with buffered formaldehyde sublimate, Zenker's fluid and Helly's fluid fixed tissues. All these are mercuric chloride containing fixatives and mercuric iron in the fixative solution can become deposited and remain bound in the tissues even after vigorous washing. Mercuric chloride fixatives were used extensively in the 1960s and 1970s in secondary post-fixation regimes. Magnesium directly inhibits Taq DNA polymerase activity if present in high concentrations. As a result, mercuric chloride fixed tissues consistently give negative results with PCR.

In terms of retrospective analysis, 40 year old material has been successfully employed to amplify specific gene sequences, however it has been reported that the size of DNA fragments prepared from samples 4 to 6 years old was often much smaller than from comparable samples which were 2 years old or less.

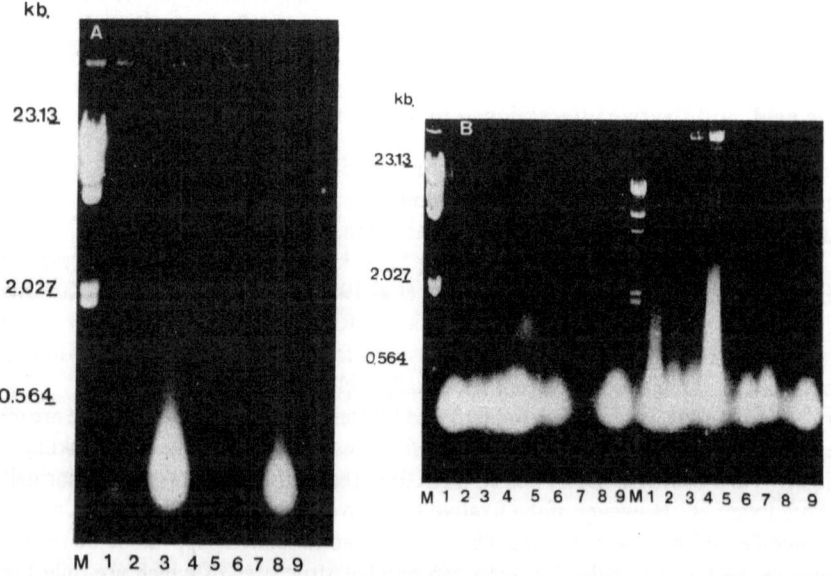

Figure 1. A: 0.7% agarose gel electrophoresis - fixed, paraffin wax embedded tissue: boiled extracted samples. M = X174 Hae III fragments. **B:** 0.7% agarose gel electrophoresis - fixed, paraffin wax embedded tissue: proteinase K extracted samples. M = X174 Hae III fragments.

NUCLEIC ACID EXTRACTION PROCEDURES

As we have seen, the success of PCR with fixed tissues is determined largely by the degree of nucleic acid/fixative interaction. However, much depends on the nucleic acid extraction technique utilised also. In our own department, we use simple boiling and proteinase K extraction for 5 days at 37°C to achieve optimum results. In general, boiled extracted samples yield low molecular weight and relatively small quantities of DNA (Figure 1A). In contrast, proteinase K yields reasonably high molecular weight DNA (500bp) (Fig 1B). Jackson et al., 1990, have shown that a 5 day proteinase K technique yields approximately 20 fold more DNA from paraffin wax embedded material than simple boiling alone. Although this is true, purification of the proteinase K extracted samples with phenol chloroform and isoamyl alcohol is required for efficient PCR. Some centres however use proteinase K extracts and then boil the samples at 100°C for 5 minutes prior to use. Even after such complicated extraction procedures, DNA may still fail to amplify. Simple techniques, such as dilution of the sample with HPLC water may help to overcome amplification failure problems. The use of special clean up kits (Iso-Gene) is also advised. Where possible, it is advisable to avoid high speed centrifugation steps, as the already fragile DNA from paraffin wax embedded material may sheer during such centrifugation steps.

As can be seen, many variables will effect the suitability of paraffin wax embedded material for use in PCR, therefore close attention must be paid to the type of fixative, fixation condition and subsequent nucleic acid extraction procedure that is applied to such material, if successful and reproducible results are required.

IN-SITU PCR AND FIXED MATERIAL

The idea of in-situ PCR has captured the imagination of many investigators. Preliminary results were reported by Bagasra et al (1990), describing a method in which a PCR was carried out on a population of cells which were then immediately subjected to in-situ hybridisation using conventional conditions. The current drawback with solution phase PCR, is that the reaction cannot be specifically linked to one particular cell type in a tissue, rather than to the tissue as a whole.

In theory, a fixed cell should act like a sponge or at least a semi-permeable dialysis bag. Cells can be fixed in ethanol or formaldehyde fixatives and they are then rendered semi-permeable. It should be possible for PCR reagents (Taq, primers, etc) to diffuse through the cell membrane and into the cytosolic and nuclear components of the cell. From the provisional results reported by Bagasra, it appeared that the majority of the amplified product appeared to remain within the nuclear or cytoplasmic component of the cell, although some products did eventually escape into the surrounding medium. Hybridisation of the amplified DNA with a suitably labelled DNA probe was then required immediately, before the amplified DNA had diffused from its site of amplification.

Nuovo et al (1991a and b, 1992a and b), Haase et al (1990), have described a technique for paraffin embedded tissues placed on glass slides to allow for the study of archival histological material using in-situ PCR. In this method, tissue sections are floated onto organosilanecoated glass slides. The tissue section is then covered with between 10-25 microlitre of amplification reagents and this is then covered by a polypropylene coverslip. A hot start PCR is performed and once the glass slide has reached 60°C-80°C, the Taq DNA polymerase is added to the slide and the coverslip is replaced. The coverslip is then anchored to the slide with nail polish and is overlaid with mineral oil to prevent drying. Target DNA denaturation is achieved at 94°C for 3 minutes with a subsequent cycling protocol of 55°C for 2 minutes and 94°C for 1 minute. PCR amplified product is subsequently detected by non-isotopic in-situ hybridisation or by direct incorporation of biotin or digoxigenin labelled nucleotides into the PCR product. Typically, 30-40 cycles of amplification are performed, but only 20-25 cycles are required, if direct incorporation of labelled nucleotide is employed. Figure 2 represents a schematic diagram of in-situ PCR.

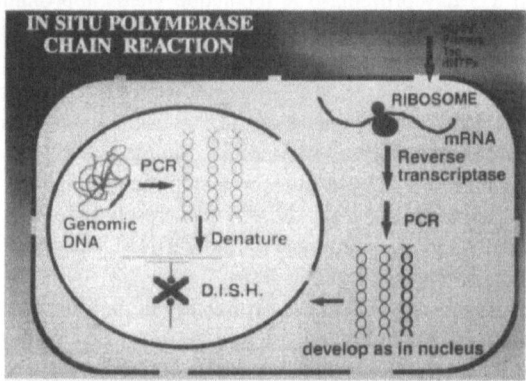

Figure 2. Schematic representation of in-situ PCR

The in-situ PCR technique represents an exciting development in the field of molecular biology. From its first report approximately 2 years ago, few publications have been seen in the literature detailing its routine applicability. Many problems still exist, with several centres reporting failure of the technique. Optimisation of section adhesion and optimal conditions for unmasking of nucleic acid are very similar to those problems encountered in normal in-situ hybridisation. The precise volume of PCR reactants used on each slide, I believe, is determined by the surface area of the cytological/biopsy material being examined. For large biopsies, 25-75 microlitres of PCR reactants are required. Small cell suspensions or cytological smears require between 10-15 microlitres of PCR reactants. Problems of leakage diffusion of PCR product are encountered frequently. The size of the amplicon determines whether or not leakage occurs from the nuclear to the cytoplasmic compartment. In order for the amplified product to remain localised within the nucleus, electrostatic forces or indeed the inducement of DNA cross-linking are required to maintain localisation of product. In my own experience, products of the order of 100bp tend to diffuse from the nuclear to the cytoplasmic compartment. However, some improvement can be achieved by immediately post-fixing the slide containing the amplified product with 2% paraformaldehyde. This may induce cross-linking of newly formed DNA strands and cause binding of newly synthesised DNA to histoneprotein subfractions and to the original DNA template within the nuclear compartment of the cell. Larger fragment amplicons appear to remain within the nuclear compartment, even in the absence of post-fixation.

The use of "hot start PCR" greatly facilitates in-situ PCR. In-situ PCR using one primer pair will not succeed if hot start is omitted. Additionally the use of a flexible coverslip to place on the PCR reactants is also advised. In my own experience I advocate the use of gel bond (FMC Bioproducts), which has a hydrophillic and hydrophobic side. It is advisable to place the hydrophobic side down, so as when the coverslip is lifted off to place the primers and Taq polymerase on the section (in the hot start protocol), no PCR reactants are lost or adhere inadvertently to the coverslip.

Ideally, in-situ PCR is performed in a deep welled slide as for conventional in-situ hybridisation. Once the coverslip has been replaced on the tissue section/cytology smear, sealing with nail varnish, rubber or indeed agarose, will achieve complete compartmentalisation of the PCR reaction within the deep welled slide. I also advocate the use of preheated mineral oil (80°C) to be placed on top of the coverslip. To maximise the rate of heat transfer, the glass slide containing the tissue section is placed in an aluminium foil boat, which in turn is placed on the heating block of the thermal cycler. The use of a thermocouple to achieve proper reaction temperature kinetics is also advised. If one uses a conventional thermal cycler heating block, the temperature at the block face and that of the slide may show large variations during the reaction.

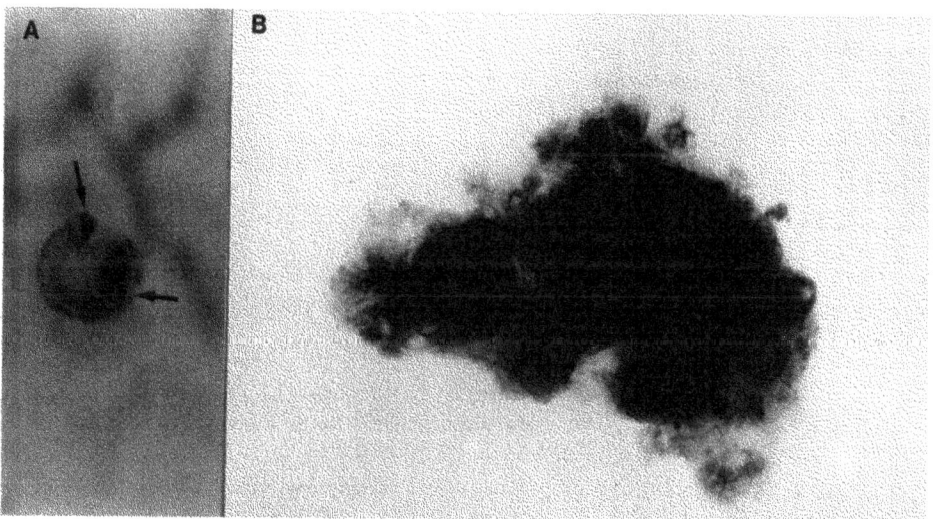

Figure 3. A: In-situ PCR of SiHa cells showing two copies of HPV 16 per cell. **B:** In-situ PCR CaSki cells showing leakage/overspill phenomenon.

Following PCR, conventional in-situ hybridisation is performed, with biotin or digoxigenin labelled internal or genomic probes. Improved performance is achieved if a labelled primer is included in the PCR reaction as we have discussed earlier. Even after rigorous precautions, patchy amplification is seen in most sections. Amplification seems to occur preferentially in some cells. This may be due to inadequate digestion during pretreatment regimes, or a lingering DNA cross-linking phenomenon due to prior fixation of cells. In addition, the thermal profile of the reaction may differ at several sites on the tissue section/cell suspension. Non-specific binding of primers must also be allowed for during the reaction. Therefore the inclusion of rigorous controls, including parallel solution phase PCR, must ideally be performed in all cases.

I have examined the in-situ PCR process as applied to fixed paraffin wax embedded cell lines. To do this, I examined SiHa cells (which contain 1-2 copies of HPV 16 per cell) and CaSki cells (containing 200-300 copies of HPV 16). Using conventional non-isotopic in-situ hybridisation, the 1-2 copies of HPV 16 per SiHa cell cannot be usually demonstrated. Figure 3A demonstrates the two copies (seen as two black dots in the nucleus of the SiHa cell). With moderate copy number viral infection (eg CaSki cell) diffusion and leakage of PCR product is encountered (Figure 3B). It appears clear from this work that in in-situ PCR, amplification of the order 100-200 fold occurs with 30-40 cycles of amplification. Ideally, the technique as currently constituted is ideal for detecting low-copy viral infection as applied to the "HPV model system". With moderate to high copy number viral infection, "leaking" phenomena predominate, which obscures cellular morphology and confounds precise localisation of the site of amplification.

In summary, I believe that this technique is not routinely applicable for clinical diagnostic purposes currently. In the future, in-situ PCR may be used for the detection of low copy number viral infections and to assess the expression of structural genes and/or viruses in biopsy/cytology smears.

ACKNOWLEDGEMENT

I wish to thank Miss Juliet Hamblin for typing the manuscript.

REFERENCES

An SF, Fleming KA. Removal of inhibitor (s) of the polymerase chain reaction from formalin fixed, paraffin wax embedded tissues. J Clin Pathol 1991, 44: 924-927

Bagasra O, Hauptman ST, Harold DO, Lischner W, Sachs M, Pomerantz R. Detection of human immunodeficiency virus type I pro-virus in mononuclear cells by in-situ polymerase chain reaction. N Engl J Med 1992, 326: 1385-1391

Barton-Rogers B, Alpart LC, Hine EAS and Bussone GJ. Analysis of DNA in fresh and fixed tissue by the polymerase chain reaction. Am J Pathol 1990, 136: 541-548

Battifora H, Kopinski M. The influence of proteinase digestion and duration of fixation on the immunostaining of keratins. A comparison of formalin and ethanol fixation. J Histochem Cytochem 1986, 44: 1095

Greer CE, Pattison SL, Kiviat NB, Manos M. PCR amplification from paraffin embedded tissues: effects of fixative and fixation time. Am J Clin Pathol 1991, 95: 117-124

Haase AT, Retzel E, Staskus K. Amplification and detection of lenti viral DNA inside cells. Proc Natl Acad Sci USA 1990, 87: 4971-4975

Jackson DP, Lewis FA, Taylor GR, Boylston AW, Quirke P. Tissue extraction of DNA and RNA and analysis by the polymerase chain reaction. J Clin Pathol 1990, 43: 499-504

Nuovo GJ, MacConnell P, Ford A, Delvenne P. Detection of human papillomavirus DNA in formalin fixed tissues by in-situ hybridisation after amplification by polymerase chain reaction. Am J Pathol 1991a, 139: 847-854

Nuovo GJ, Gallery F, MacConnell P, Becker J, Bloch W. An improved technique for the in-situ detection of DNA after polymerase chain reaction amplification. Am J Pathol 1991b, 139: 1239-1244

Nuovo GJ. In-situ detection of PCR amplified DNA and cDNA. Amplifications: a forum for PCR users. 1992a, 8: 1-3

Nuovo GJ, Becker J, Margiotta M, MacConnell P, Comite S, Hochman H. Histological distribution of polymerase chain reaction amplified human papillomavirus 6 and 11 DNA in penile lesions. Am J Surg Pathol 1992b, 16: 269-275

McGhee JD, Von Hippel PH. Formaldehyde as a probe of DNA structure 1. Reaction with exocyclic and minor groups of DNA bases. Biochem 1975, 14: 1281-1295

McGhee JD, Von Hippel PH. Formaldehyde as a probe of DNA structure 2. Reaction with endocyclic imino groups of DNA bases. Biochem 1975, 14: 1297-1303

McGhee JD, Von Hippel PH. Formaldehyde as a probe of DNA structure 3. Mechanism of the initial reaction of formaldehyde with DNA. Biochem 1976, 16: 3267-3276

McGhee JD, Von Hippel PH. Formaldehyde as a probe of DNA structure 4. Equilibrium, denaturation of DNA and synthetic polynucleotides. Biochem 1976, 16: 3276-3293.

IMPACT OF PCR ON THE PATHOLOGIST'S WORLD

D. Myerson

Fred Hutchinson Cancer Research Center, Seattle, USA

PCR has burst upon the pathologist's world as both a promise and a peril. There are many pathology laboratories and research centers throughout the world that perform PCR, many in a more or less routine manner to detect a variety of conditions. They can be divided into 3 general categories: I) Low copy number of foreign nucleic acid. These are generally infectious diseases. Two examples are Hepatitis C Virus (HCV) and Cytomegalovirus (CMV). II) Single copy per genome. These usually involve genetic variation, either mutations or polymorphisms. Two examples are the human histocompatibility locus (HLA) and cystic fibrosis. III) Low copy number of human DNA. This category includes the detection of malignancy, or the associated chromosomal translocations or aberrant expression, either in quantitatively or qualitatively. Examples are the bcr-abl translocation associated with chronic myelogenous leukemia and the increased expression of her2/neu seen in some breast carcinomas. The first two categories will be considered in this synopsis.

The main considerations in developing a PCR test for clinical use is the need for simplicity, reproducibility, and lack of contamination. PCR intrinsically is not a simple procedure. My laboratory is performing routine assays for hepatitis C virus in bone marrow transplant recipients in a clinical research setting. We have chosen to modify the research procedures only minimally (Figure 1). This yields what probably is the most complicated test in all of clinical pathology. RNA must be laboriously extracted from serum or plasma. For this we use the usual guanidinium hydrochloride solubilization, and phenol-chloroform-ethanol extraction procedures. To minimize false positives and false negatives, all samples are done at least in duplicate. Negative controls are randomly interspersed, averaging every 6th tube.

The HCV PCR utilizes a second amplification with concentric primers as originally described by Okamoto et al (1990). The potential for contamination is increased because the reaction tube must be opened after the reverse transcriptase step and again after the first PCR step. Careful titration of these steps has lead to a procedure in which contamination is rare. It was found necessary to carry out the first outer primer PCR for 30 steps. This was determined by the number of steps which reproducibly was required to amplify 10 copies of DNA. The inner primers are cycled for 20 cycles. This was determined by arbitrarily choosing 10,000 copies as the sensitivity desired. It was assumed that if 10,000 copies were not produced after 30 cycles of the first PCR, the result was negative. Additionally, to ensure specificity, the product was detected by an internal digoxigenin-labelled 145 base pair PCR synthesized probe. Therefore,

Methods in DNA Amplification, Edited by
A. Rolfs *et al.*, Plenum Press, New York, 1994

the test was designed to yield either a strongly positive or a strongly negative signal. It was made intentionally made non-quantitative. In so doing, the problem of clinical interpretation comes to the fore.

The PCR was exclusively used for accurate assessment of HCV status in marrow transplant patients. Marrow transplant patients have no functioning immune system until engraftment of the donor marrow and often remain on long term immunosuppressive therapy for treatment of graft-versus-host disease. Additionally they often receive immune globulin, which contains exogenous antibodies to HCV which would make the usual antibody test uninterpretable. Therefore, in order to assess the frequency of hepatitis C virus transmission and its natural history, a direct measurement of the virus is necessary rather than the usual antibody testing. PCR was employed for this purpose.

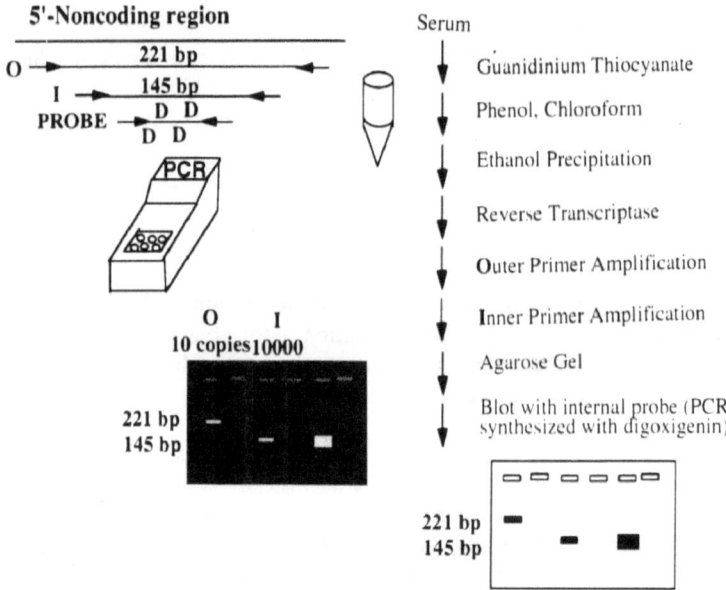

Figure 1. Diagram of the PCR procedure for HCV.

Does the infusion of hepatitis C positive bone marrow into bone marrow transplant recipients result in viral transmission? Patients were studied who were HCV negative pretransplant and received marrow from donors having antibodies to HCV by the first generation ELISA test. Each donor was also tested by PCR. After marrow transplantation, serial serum specimens were obtained from these patients and analyzed for HCV RNA by PCR. There was excellent correlation between HCV detection by PCR in the donor and acquisition of HCV by the recipient. This suggests that PCR is a good method to predict whether a seropositive donor will transmit the virus.

Does the transfusion of hepatitis C positive blood products (platelets) result in transmission of HCV? Transmission of HCV by blood products was also studied in a similar manner. PCR accurately predicted the infectiousness of blood products.

In these, relatively small studies, PCR was generally neither too sensitive, detecting positive results in patients' tissue who don't ultimately transmit the virus nor was it too insensitive, not detecting virus in patients serum whose bone marrow ultimately does transmit the virus.

In these sorts of studies, the issue of sensitivity is paramount. PCR has the capability of detecting a single genome copy. Unfortunately, in the case of RNA detection by PCR, the reverse transcriptase step only has an efficiency of about 10%. The ability to detect a single HCV particle is therefore not always possible. The volume of serum tested and the sensitivity of the PCR may affect the clinical result. The gold standard in this case is the transmission of hepatitis C virus by the actual marrow or blood product infusion.

The clinical interpretation of a PCR result is often problematical. It is frequently not enough to determine whether a person has been or has not been infected with a particular infectious agent. It may be necessary to somehow predict something about the course of the infection, whether a certain bodily fluid might be infectious, or whether a particular infection will respond to antibiotics. Since unmodified PCR is intrinsically not a quantitative technique, the result of a PCR may vary, depending on small differences in the sensitivity of a particular assay. Quantitative PCR, or PCR designed to detect a specific genomic transcript, may be necessary for many clinical purposes. For example, the amount of HCV RNA may be correlated to the degree of HCV induced hepatitis. The difficulty of interpreting a positive PCR result is one which is often not considered in a research setting but is critical to consider in a clinical setting. This principle is further illuminated by our study with CMV.

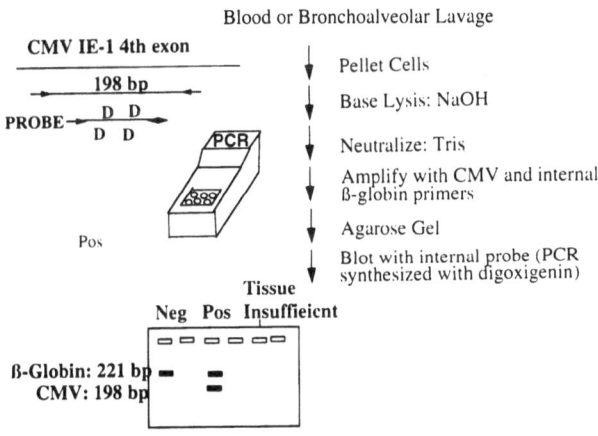

Figure 2. Diagram of the procedure for HCV

A multiplex PCR procedure was developed, able to detect both CMV and assay the adequacy of the sample via additional human ß-globin primers (Myerson et al 1993) (Figure 2). Since CMV is a DNA virus, we used a simple base lysis procedure to prepare the sample. A single primer pair was used to detect CMV. The resulting gel sometimes showed a light band, and sometimes an extraneous band at approximately the proper location on the gel. To unequivocally determine whether the result was positive or negative, it was necessary to perform a Southern blot. We probed the Southern blot with a digoxigenin labeled probe synthesized by PCR. It is important that the probe be completely internal to the primers employed in the PCR, to avoid detection of non-specifically extended primers. However, we did not rely on the sensitivity of the Southern blot for the sensitivity of the reaction. It is almost always preferable to increase sensitivity by applying better conditions, more cycles, or different primers in the PCR reaction than it is to increase the sensitivity in a way prone to background, such as by using very hot primers in a Southern blot.

Southern blots are generally not highly quantitative. This is especially true with the short pieces of DNA to be detected, such as 200 base pair PCR products. These products stick poorly to many of the blotting membranes. Using a charged Nylon membrane, several membranes were stacked on top of each other and the gel blotted with paper towels as usual. Product was found in all the membranes.

We compared the detection of CMV in the blood to other tests, and the presence of serious CMV disease, in this case, CMV interstitial pneumonia. At present, it is not certain whether the PCR assay will have any impact on the clinical diagnosis of CMV. It appears to be too sensitive in detecting serious CMV disease, but it may have applicability in determining how long treatment is to be given. A randomized clinical trial is in order. For the test to be reproducibly performed between the various labs, careful quality control is essential to make sure the tests are equally sensitive. Simple detection of an infectious agent in the most sensitive possible manner may not be what would yield the most useful clinical information in this case.

HIV presents a similar quandary. To simply diagnose infection, the simple presence of the virus is a sufficient result. However, as effective treatments are being developed and patients are living longer, it may be necessary to ascertain the state or quantity of the virus, or what particular cells are infected.

The second category of PCR in the pathologist's world is in the diagnosis of genetic mutations or diseases. Here, the target is present at one or two copies per genome, much more concentrated than an extraneous piece of nucleic acid such as in infectious disease. Sensitivity, therefore, is not generally a difficult problem. The problem is often obtaining sufficient knowledge about the potential defects to make an adequate identification.

The determination of human histocompatibility types is currently being performed in large part by PCR at the Fred Hutchinson Cancer Research Center and many other locations. There are six HLA loci- A, B, C, DP, DO and DR. Each has many alleles. The main research and clinical problem is to determine which sequences correlate to graft rejection when they differ in the host and the grafted tissue. The HLA loci have been serotyped, so the problem often resolves to how the sequences relate to the serotype or mixed lymphocyte cytotoxity cell typing.

To type an HLA locus, the DR locus for example, it is necessary to find a suitably conserved portions of the gene so that a single set of primers will amplify all variants. The variable sequences at issue must be internal to the primers. For example, a portion of the DR-B1 gene is used. After amplification, it is then necessary to determine the sequence of the gene. To do this, single strand oglionucleotide probes (SSOP) of the many possible sequences are hybridized to the PCR product in a dot blot performed under exacting hybridization conditions. Depending on the result, specific additional pair(s) of "group specific" primers must then be applied. The resultant product is again identified by SSOP analysis, this time with group specific probes.

Each SSOP pattern of hybridization corresponds to an HLA-DR allele.

Similar to the identification of alleles, PCR may used to identify genetic mutations. Often the possible mutations are spread over a very large area of DNA. One needed assay is that for cystic fibrosis. CF is an autosomal recessive disorder with approximately 1 out of 20 Caucasians being unaffected carriers of the gene. It is the most common genetic disease in that population, affecting 1 out of every 2000 births.Fortunately, about 75% of the mutant genes are due to a uniform deletion of a single trinucleotide-the ΔF506 mutation. This is easy to detect using PCR amplified material.

Unfortunately, however, the remaining mutations are distributed rather evenly over more than one hundred different mutations, each present at a low frequency. Testing for the ΔF503 mutation only, 75% of the carriers will be detected. Testing for the 6 most common mutations, 35% will be detected. And testing for the 20 most common mutations, 87% will be detected.

The CF gene is a huge one, spanning 250kb of DNA, and consisting of 27 exons. It codes for the cystic fibrosis transmembrane conductance regulator. Therefore, in order to detect all cystic fibrosis mutations, many different regions of the genome must be amplified and many different detection probes must be synthesized. This has resulted in great difficulty in using PCR to perform this assay with high sensitivity, but relatively ease in performing it with low sensitivity. Assaying only for ΔF503, only 50% of all couples at risk will be identified. At present, with a detection sensitivity of 65%, 72% of all couples at risk can be identified.

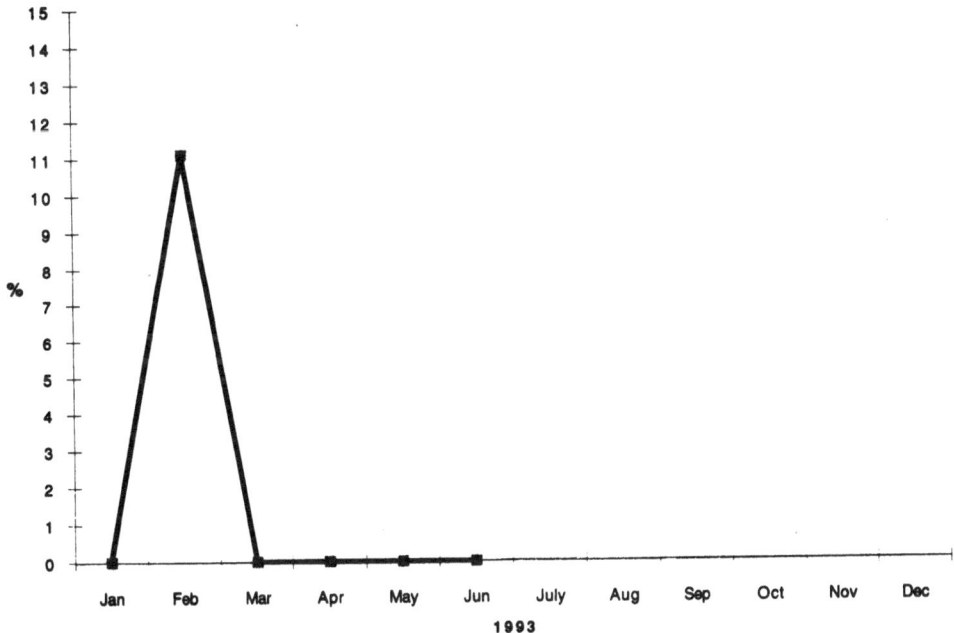

Figure 3. Example of quality control record showing the frequency of negative control that were falsely positive. The peak occurred when a new technician entered the laboratory. The percent of negative controls that were positive (by month) is given.

One last topic is that of quality control (Figure 3). Many studies have shown that different laboratories may produce widely differing results using purportedly the same PCR procedure. PCR must be quality controlled in a similar fashion to any clinical laboratory procedure. Records must be kept of the false positive and false negative rates, and acceptable limits must be ascertained. It is not reasonable to simply throw away the results of a run with a sporadic false positive, but instead an effort must be made to keep the false positive rate very low, so that all the results may gain a certain confidence level. In practice duplicate samples are used, and only a low frequency of falsely positive or falsely negative controls is tolerated. The necessity for strong quality control is magnified by the extreme sensitivity of PCR rather than minimized.

As PCR moves from the research laboratory to the clinical research laboratory to the clinical pathologist's laboratory, some technological advances that would facilitate the transit. There are several items on the pathologist's wish list.

The first item is simple and rapid sample preparation. The simplest preparation may be merely putting the blood or tissue into the reaction. Unfortunately, the DNA is often not accessible or inhibitors may be present. The traditional DNA and RNA extractions yield a very laborious procedure. Rapid preparation, especially for RNA, would be a major advance. The easy use of large sample volumes to improve sensitivity, would be a further improvement.

A second advance would be the use of a single tube for RT-PCR without reopening it. There are some high temperature polymerases, such as Tth, which can also reverse transcribe, and is hoped that such a procedure might be incorporated in a robust single tube assay for RNA.

The third wish is for a good, inexpensive homogeneous detection system. It could be read in the very tube in which the PCR was carried out without further separation. Such homogeneous detection systems have been reported and it is hoped they will be perfected.

The fourth wish is for a contamination proof procedure in which the product cannot contaminate the input. Although somewhat laborious, such a procedure is already available through incorporation of the nucleotide dUTP in place of ^{32}P. The product incorporating the dUTP may subsequently be destroyed without destroying input DNA.

The fifth wish is for an efficient, easy to detect, quantitative PCR. Quantitative PCR has been performed by co-amplifying the sample DNA or RNA and a similar known DNA or RNA sequence, and then comparing the relative quantities of both products. A homogenous detection system able to detect and differentiate both products would be ideal.

The final wish is for an automated procedure including automated sample preparation, automated thermal cycling, and an automated detection.

In the pathologist's world there are many indirect procedures being performed to detect infectious diseases as well as genetic mutations, aberrations, and polymorphisms. PCR will obtain a place in the pathologist's world in many of these assays, and also permit the assessment of entities not presently detectable. It should not be forgotten, however, that the PCR result must lead to a relevant clinical prediction, in order to fully integrate PCR into the pathologist s world.

REFERENCES

Okamoto H, Okada S, Sugiyama Y, Tanaka T, Sugai Y, Akahane Y, Machida A, Mishiro S, Yoshizawa H, Miyakawa Y. Detection of hepatitis C virus RNA by a two-stage polymerase chain reaction with two pairs of primers deduced from the 5′- noncoding region. Japan J Exp Med 1990, 60:215-222

Myerson D, Lingenfelter PA, Gleaves CA, Meyers JD, Bowden RA. Diagnosis of cytomegalovirus pneumonia by the polymerase chain reaction with archived frozen lung tissue and bronchoalveolar lavage fluid. Am J Clin Pathol 1993, 100: 407-413

SENSITIVE AND RAPID DETECTION AND QUANTIFICATION OF NUCLEIC ACIDS

L. Cross, C. Potts and J. G. Anson

Amersham International plc, Cardiff Laboratories, Forest Farm, Whitchurch, Cardiff, Great Britain

INTRODUCTION

The polymerase chain reaction (PCR; Saiki et al 1988) is increasingly being used to quantify the number of copies of nucleic acid target in a sample. PCR quantification is important for many research and clinical applications such as analysis of gene expression (Gilliand et al 1990; Murphy et al 1990; Hoof et al 1991) or monitoring viral (Kellog et al 1990; Holodniy et. al 1991; Menzo et al 1992; Bavin et al 1993) and bacterial (Leigh et al 1993) infections. Due to the sensitivity of PCR, small variations in the reaction efficiency will result in significant differences in the amount of final product formed. A number of strategies have been developed which can minimize the variability of the reaction and hence increase the robustness of using PCR as a quantitative technique (Ferre 1992).

A common feature of all quantitative PCR approaches is the need to determine the amount of PCR product formed at the end of the reaction. The sensitivity and accuracy of this step is critical, and generally involves relatively complex and time-consuming manual operations. Two examples of commonly used procedures for end-point quantification are hybridization of products to radiolabelled probes (Dickover, 1990), and incorporation of radiolabel during the PCR and excision of the relevant bands from gels followed by scintillation counting (Siebert and Larrick 1993).

We have developed an end-point assay which combines the convenience and speed of Scintillation Proximity Assay (SPA; Bosworth and Towers, 1989) with the power of PCR to provide a novel assay system for the quantification of amplified nucleic acid molecules. SPA is a unique technology which relies on the use of fluomicrospheres (SPA beads) coated with acceptor molecules which are capable of binding radiolabelled ligands in solution. The technique is performed in assay buffer, and requires the use of a labelled ligand with an isotope that emits low-energy radiation (eg. [³H]), which is dissipated easily into an aqueous medium. Labelled ligands which can specifically bind to the beads will be in close enough proximity to

Methods in DNA Amplification, Edited by
A. Rolfs *et al.*, Plenum Press, New York, 1994

activate the fluor and produce light which is detectable by a scintillation counter. Due to the aqueous environment, the majority of unbound labelled ligands are too far from the beads to enable the transfer of energy. This characteristic of SPA removes the need for any separation steps, making the assay homogeneous.

ASSAY PROCEDURE

We have developed two assay formats which use SPA beads to quantify the amount of PCR product formed. In the first format, one of the primers is biotinylated, and tritiated nucleotides are incorporated during amplification. At the end of the reaction, an aliquot of the amplified DNA, which will be biotinylated and tritiated, is added to streptavidin-coated SPA beads. Only biotinylated products will bind to the beads, and the amount of product bound can be quantified by direct scintillation counting. This removes the need to add scintillation cocktail and reduces the problem of toxic organic waste disposal, as the SPA beads are non-toxic. Direct quantification of the PCR using SPA also removes the requirement to analyze the reaction products on agarose gels, and then to generate quantitative data relating to the amount of DNA present in the band. This significantly reduces the amount of time required between completion of the PCR to generation of data on the amount of DNA produced. The assay format allows multiple samples to be processed extremely rapidly; 20 samples can be quantified in less than half an hour. An example of the data produced using this assay format is given in figure 1.

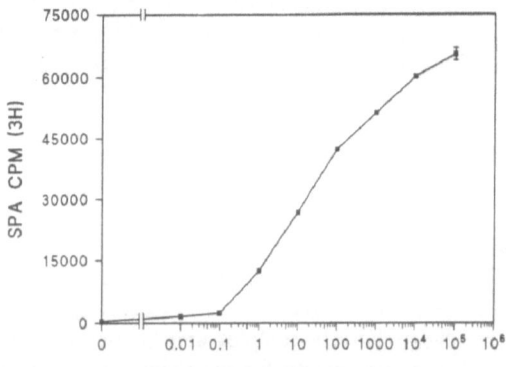

Amount of initial pKV460 target (fg)

Figure 1. Typical PCR standard curve generated using the SPA assay system. A dilution series of plasmid pKV460 was amplified using primers CAT1 and CAT13 (biotinylated). After 28 cycles of amplification by PCR, streptavidin-coated SPA beads were added to an aliquot of each reaction and quantified by scintillation counting. Using the assay a linear range over 6 orders of magnitude was obtained, with a sensitivity down to less than 10 copies of original target DNA.

This assay format requires a high degree of specificity from the PCR, as all products which are biotinylated and tritiated will contribute to the overall signal obtained. For situations where an increased level of specificity is required, an alternative approach can be used. PCR products are generated using standard non-biotinylated primers, and tritium is incorporated during amplification as in the previous format. On completion of the reaction, the double stranded products are denatured at 95°C in the presence of a biotinylated oligo which is specific for the PCR product of interest. The biotinylated oligo is then allowed to anneal to the specific DNA strand, and the product of this annealing reaction can be captured on SPA beads and quantified by scintillation counting (figure 2). In addition to enhanced specificity, this assay format also allows more than one PCR product to be quantified from a single reaction. This can be performed by dividing the reaction on completion, and annealing individual specific biotinylated oligos in separate reactions.

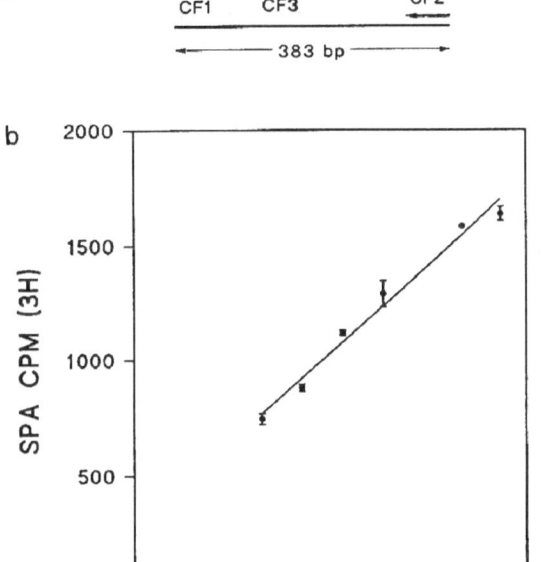

Figure 2. Demonstration of the use of the oligo capture approach. (a) Marker region of the human cystic fibrosis gene containing a 383 bp amplicon flanked by primers CF1 and CF2, and showing the position of biotinylated capture oligo CF3. (b) Standard curve generated using the oligo capture procedure. Human genomic DNA was serially diluted and amplified by PCR using primers CF1 and CF2. Following amplification, the PCR products were denatured by heat and the oligo CF3 added and allowed to anneal to the complementary DNA product. Streptavidin-coated SPA beads were added and amount of product quantified by scintillation counting. Using the procedure described, the assay is linear over 2 orders of magnitude of target DNA

APPLICATION OF SPA IN QUANTITATIVE PCR

Semi-quantitation

The relationship between the initial amount of target present (A) and the amount of DNA produced after PCR (Y) can be expressed as $Yn = A(1+R)^n$, where R is the efficiency of the PCR and n is the number of cycles (Chelly et. al., 1988). Small variations in the reaction efficiency therefore translate into large variations in the amount of product generated at the end of the reaction.   For quantitation to be accurate, the reaction should be controlled to ensure maximum possible efficiency and terminated whilst in the exponential phase.   In practice this requires that the PCR conditions are carefully optimized, and that controls are chosen which can be used to standardize the reaction.   This approach has been used to quantitate levels of gene expression by RT-PCR (Kinoshita et al 1992; Melby et al 1993) and for viral quantitation (Yang et al 1993).

The SPA approach using a biotinylated primer is a simple and sensitive format which can be adapted for use with semi-quantitative strategies. The PCR needs to be optimized to ensure that the reaction is stopped and products quantified during the exponential phase. A standard curve is constructed from serial dilutions of a known amount target, and test samples quantified from extrapolation of the curve. Samples can also be standardized against a control target sequence (eg. ß-actin or glyceraldehyde 3-phosphate dehydrogenase for gene expression studies) in a separate PCR reaction. If standardization of the individual reactions is required, primer sets for both the target of interest and the internal control can be used in the same PCR. In this case, the third oligo approach employing a specific biotinylated oligo to anneal to the PCR products can be used to determine the relative amounts of each product formed at the end of the PCR, by using separate biotinylated oligos specific for the target and the control.

Competitive PCR

This approach is more applicable when absolute quantification is required. The method uses an internal control DNA or RNA template which has the same primer binding sites as the target sequence of interest. Reactions are spiked with known amounts of the control, and at the end of the PCR the two sets of products generated (control and target) are quantified. When the final amount of each product is the same, the assumption is that there was an equivalent amount of both control and target PCR template at the start of the reaction. The initial amount of target of interest can therefore be deduced. Because control and target are amplified together, there is no requirement to ensure that the reaction is terminated in the exponential phase, and therefore optimization is not critical in contrast to the semi-quantitative approach.

Approaches to competitive PCR differ in the methods used to discriminate between the products generated from the target and those from the control. Commonly the control is designed to produce a modified product which can be differentiated by either size or internal sequence (Wang et al 1989; Piatak et al 1993). Alternatively, a unique restriction site can be engineered into the control amplimer and subsequently used to differentiate between products generated from control and target sequences (Becker-Andre and Halbrook 1989; Stieger et al 1991; Fox et al 1992).

The SPA-based assay described can be readily adapted to quantify products from competitive PCR. The approach used will depend on the method chosen to discriminate between the control and target products. If the control and target PCR product each have a region of unique internal sequence, individual biotinylated oligos specific for each product can be designed to differentiate and quantify after competitive PCR. A single set of non-biotinylated primers can be used to amplify both target and variable known amounts of control in a competitive PCR. At the completion of the reactions, the samples can be divided into two separate sets of tubes.   The biotinylated target-specific oligos are annealed to the PCR products in one set, and the biotinylated control-specific oligos are used in the second set.

SPA beads are then added to each annealing reaction to quantify the amount of target-specific and control-specific product formed in each individual competitive PCR.

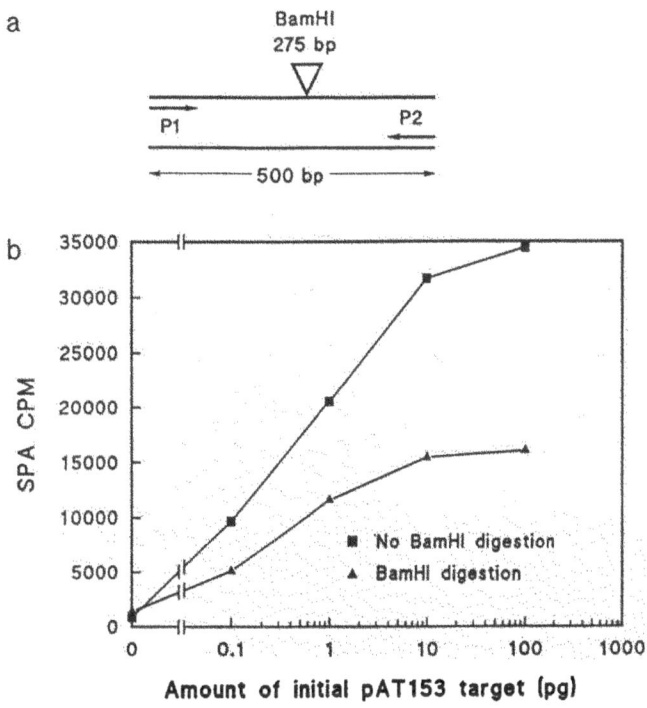

Figure 3. SPA assay format for detection of a restriction site in amplified DNA. (a) Portion of plasmid pAT153 containing a 500 bp amplicon flanked by specific PCR primers P1 and P2 (biotinylated). The position of the unique *Bam*HI site is indicated. (b) The PCR was performed on serially diluted pAT153 template DNA. Post-amplification an aliquot of the reaction was digested to completion with *Bam*HI. Digested and undigested PCR products were added to streptavidin-coated SPA beads and quantified by scintillation counting. The digested PCR products gave approximately half the signal of the corresponding undigested sample, as predicted by the position of the restriction site in the amplicon. The ratio of digested:undigested products can therefore be deduced by the SPA assay.

If the control amplimer has a unique restriction site, a single biotinylated primer can be used in the competitive PCR amplification reaction. After the PCR, aliquots of the samples are digested with the restriction enzyme, and then both digested and undigested samples can be quantified using SPA beads. This is possible as SPA can be used to detect a reduction in SPA signal associated with a reduction in the size of the product caused by restriction digestion (figure 3). The ratio of control (cut) amplimer to target (uncut) in the aliquot digested with enzyme can be determined by the degree of signal reduction observed compared to the undigested aliquot. This relationship can be used to identify the samples containing equivalent amounts of control and target DNA.

CONCLUSIONS

The method for PCR product quantification reported provides a rapid and convenient alternative to traditional approaches involving separation of products on agarose gels and quantification following isotopic or non-isotopic detection. The SPA assay is compatible with a number of commonly used approaches to quantitative PCR, where it is useful as an end-point assay. The only divergences required from the standard PCR is the incorporation of a tritiated nucleotide tracer during the amplification reaction, and the use of either a biotinylated primer or capture oligo. The biotinylation of oligos can be performed during automated synthesis using commercially available phosphoramidites.

As the SPA assay is performed downstream of the amplification reaction it is therefore independent of the procedure used. Thus the principle of the SPA assay can be applied to alternative amplification techniques.

REFERENCES

Bavin PJ, Giles JA, Deery J, Crow PD, Griffith PD, Emery VC and Walker PG. Use of semi-quantitative PCR for human papillomavirus DNA type 16 to identify women with high grade cervical disease in a population presenting with a mildly dyskaryotic smear report. Br. J. Cancer 1993, 67: 602-5.

Becker-Andre M and Hahlbrock K. Absolute mRNA quantification using the polymerase chain reaction (PCR). A novel approach by a PCR aided transcript titration assay (PATTY). Nucl. Acids Res. 1989, 17: 9437-46.

Bosworth N and Towers P. Scintillation proximity assay. Nature 1989, 341: 167-68.

Chelly J, Kaplan J-C, Marie P, Gautron S and Kahn A. Transcription of the dystrophin gene in human muscle and non-muscle tissues. Nature 1988, 333: 858-60.

Dickover RE, Donovan RM, Goldstein E, Dandekar, Bush CE and Carlson JR. Quantitation of human immunodificiency virus DNA by using the polymerase chain reaction. J. Clin. Microbiol. 1990: 2130-33·

Ferre F. Quantitative or semi-quantitative PCR: reality versus myth. PCR Methods Applic. 1992, 2:1-9.

Fox JC, Griffith PD and Emery VC. Quantification of human cytomeglavirus DNA using the polymerase chain reaction. J. Gen. Virol. 1992, 73: 2405-8.

Gilliand G, Perrin S, Blanchard K and Bunn HF. Analysis of cytokine mRNA and DNA: detection and quantitation by competitive polymerase chain reaction. Proc. Natl. Acad. Sci. USA 1990, 87: 2725-29.

Holodniy M, Katzenstein DA, Sengupta S, Wang AM, Casipit C, Schwartz DH, Konrad M, Groves E and Merigan TC. Detection and quantification of human immunodeficiency virus RNA in patient serum by use of the polymerase chain reaction. J. Infec. Dis. 1991, 163: 862-66.

Hoof T, Riordan JR and Tummler B. Quantitation of mRNA by the kinetic polymerase chain reaction assay: a tool for monitoring P-glycoprotein gene expression. Anal. Biochem. 1991, 196: 161-69.

Kellog DE, Sninsky JJ and Kwok S. Quantitation of HIV-1 proviral DNA relative to cellular DNA by the polymerase chain reaction. Anal. Biochem. 1990, 189: 202-8.

Kinoshita T, Imamura J, Nagai H and Shimotohno K. Quantification of gene expression over a wide range by the polymerase chain reaction. Anal. Biochem. 1992, 206: 231-35.

Leigh TR, Gazzard BG, Rowbottom A and Collins JV. Quantitative and qualitative comparison of DNA amplification by PCR with immunofluorescence staining for diagnosis of *Pneumocystis carinii* pneumonia. J. Clin. Pathol. 1993, 46: 140-44.

Melby PC, Darnell BJ and Tyron VV. Quantitative measurement of human cytokine gene expression by polymerase chain reaction. J. Immunol. Methods 1993, 159: 235-44.

Menzo S, Bagnarelli P, Giacca M, Manzin A, Varaldo PE and Clementi M. Absolute quantitation of viremia in human immunodificiency virus infection by competitive reverse transcription and polymerase chain reaction. J. Clin. Microbiol. 1992, 30: 1752-57.

Murphy LD, Herzog CE, Rudick JB, Fojo AT and Bates SE. Use of polymerase chain reaction in the quantitation of *mdr*-1 gene expression. Biochemistry 1990, 29: 10351-56.

Piatak M, Luk K, Williams B and Lifson JD. Quantitative competitive polymerase chain reaction for accurate quantitation of HIV DNA and RNA species. BioTechniques 1993, 14: 70-80.

Platzer C, Richter G, Uberla K, Muller W, Blocker H, Diamantstein T and Blankenstein T. Analysis of cytokine mRNA levels in interleukin-4-transgenic mice by quantitative polymerase chain reaction. Eur. J. Immunol. 1992, 22: 1179-84.

Saiki RK, Gelfand DH, Stoffel S, Scharf SJ, Higuchi R, Horn GT, Mullis KB and Erlich HA. Primer-directed enzymatic amplification of DNA with a thermostable DNA polymerase. Science 1988, 239: 487-91.

Siebert PD and Larrick JW. PCR MIMICS: competitive DNA fragments for use as internal standards in quantitiative PCR. BioTechniques 1993, 14: 244-49.

Stieger M, Demolliere C, Ahlborn-Laake L,and Mous J. Competitive polymerase chain reaction assay for quantitation of HIV-1 DNA and RNA. J. Virol. Methods 1991, 34: 149-60.

Wang AM, Doyle MV and Mark D. Quantitation of mRNA by the polymerase chain reaction. Proc. Natl. Acad. Sci. USA 1989, 86: 9717-21.

Yang B, Yolken R and Viscidi R. Quantitative polymerase chain reaction by monitoring enzyme activity of DNA polymerase. Anal. Biochem. 1993, 208: 110-16.

POLYMORPHIC KERATINS AS DETECTED BY PCR AND SSCP

D. Mischke[1], R. Wanner[1] and B. P. Korge[2]

[1]Institut für Experimentelle Onkologie und Transplantationsmedizin, Universitätsklinikum Rudolf Virchow, Freie Universität Berlin, Berlin, [2]The Skin Biology Branch National Institute of Arthritis and Musculoskeletal and Skin Diseases, National Institutes of Health, Bethesda, USA

Genetic polymorphism of keratins at the protein level due to allelic variation has been described for K1, K4, K5, and K10. In order to understand the molecular basis of the differences among the alleles of these genes, we have analyzed their N- and C-terminal domains following amplification of genomic DNA by the polymerase chain reaction. Whereas the K1 and the K10 alleles differ in size of their carboxyl-terminal V2 subdomains, the alleles of the K4 and K5 genes diverge in their amino-terminal domains.

Since there are at least 11 alleles of the K10 gene and, combined, almost as many alleles of the K1, K4, and K5 genes, the polymorphic human keratin genes by themselves are highly informative markers to elucidate the potential involvement of keratins in hereditable disorders of squamous cell differentiation using genetic linkage analyses within the two keratin gene clusters located on chromosomes 17q and 12q, respectively.

INTRODUCTION

From keratins the intermediate filaments of all epithelial cells are assembled to span the network within the cytoplasm that endows structural organization and strength to the cells.

In humans, there are more than 20 different keratins (K) encoded by at least as many differentially expressed genes. Based on their relatedness they can be subdivided into two classes: the acidic type I keratins (K9 - K20) and the neutral-basic type II keratins (K1 - K8). Within a particular epithelial tissue or cell type, specific pairs consisting of one type I and one

Methods in DNA Amplification, Edited by
A. Rolfs *et al.*, Plenum Press, New York, 1994

type II keratin are expressed according to a developmentally controlled and differentiation-specific program (Moll et al., 1982; Sun et al., 1984; Fuchs, 1988; Steinert and Roop, 1988; Oshima, 1992). Keratin genes are clustered in the human genome: all type I keratins are located on chromosome 17q11-q21 and all type II keratins on chromosome 12p12-q11 (e.g. Lessin et al., 1988; Barletta et., 1990; Rosenberg et al., 1991).

In terms of protein structure, keratins, like other intermediate filament proteins, share a central a-helical domain of approximately 310 amino acids which is flanked by non-helical amino-terminal head and carboxyl-terminal tail domains that vary considerably in size and sequence (for a review see Parry and Steinert, 1992).

Only in the last two years, several groups have identified point mutations in keratin genes causing autosomal dominant skin blistering diseases such as epidermolysis bullosa simplex or epidermolytic hyperkeratosis (Bonifas et al., 1991; Coulombe et al., 1991; Lane et al., 1992; Compton et al., 1992; Rothnagel et al., 1992; Cheng et al., 1992; Chipev et al., 1992). Interestingly, all of these mutations are missense mutations that occur predominantly at the beginning and end of the a-helical domain, suggesting that these regions are of particular functional importance (see also Hatzfeld and Weber, 1990; Herrmann et al., 1992).

Since there are other disorders of epithelial differentiation of as yet unknown etiology, polymorphic markers localized directly within the two clusters of keratin genes on chromosome 12q and 17q would be of immediate significance for linkage analyses in families affected with inherited epithelial diseases of suspected keratin abnormalities. Here we review the available data on such markers, the polymorphic human keratins K1, K4, K5, and K10.

MATERIALS AND METHODS

Preparation of nucleic acids

Genomic DNA was isolated according to standard procedures (Maniatis et al., 1982) using proteinase K digestion and phenol/chloroform extraction. The ethanol precipitated DNA was dissolved in water.

Polymerase chain reaction

PCR amplifications were performed according to the protocol supplied by Perkin-Elmer (Norwalk, CT) using, in a 30 μl reaction, 30 ng of genomic DNA and 100 nM of the appropriate oligonucleotide primers (see Table 1; sequence information from Johnson et al. (1985) for K1; Wanner et al. (1993) and Leube et al. (1988) for K4; Eckert and Rorke (1988) and Lersch and Fuchs (1988) for K5; Zhou et al. (1988) and Rieger and Franke (1988) for K10). Amplifications of K4 and K5 were performed as a Touch-down PCR with 4 cycles of 1 min 94°C, 1 min 60°C, and 1 min 72°C, 15 cycles of 1 min 94°C, 1 min 58°C, and 1 min 72°C, and 20 cycles of 1 min 94°C, 1 min 56°C, and 1 min 72°C. An elongation step of 7 min at 72°C completed the amplification procedure.

Table 1. List of oligonucleotide primers used for PCR

Denotation	Sequence (5' to 3')	Product Length (bp)
K1-2a	GATGTCTGGAGAATGTGCCCCG	641/620
K1-2b	GGCTGGGACAAATCGACCTCGG	
K4-20a	CAACCTCAGGGGGAACAAAAGC	177/135
K4-20b	CACCCTTACCACTGAAGGAGCC	
K5-11d	GTGGAGGCAGCTTCAGGAACCG	251
K5-11b	CTGGATGCTGGGGTCGATTTGC	
K10-1a	AACGGCAACTGGAAAGCTACCC	variable sizes
K10-1b	GATGAAGACTCGCCCACGGACC	

The K1 and K10 genes were amplified using reaction mixes supplemented with 10% glycerol and 3% formamide (v/v) and 38 cycles consisting of 1 min at 94°C, 2 sec at 58°C (60°C for K1), and 5 sec at 72°C in the TC1 thermal cycler with an elongation step at 72°C for 7 min (Speed-up PCR).

SSCP analysis

The K4 and K5 amplification products were used directly for SSCP analysis and 1/30th of the original PCR sample was denatured in 10 vol of 95% formamide, 10 mM NaOH, 10 mM EDTA, 0.05% Bromphenolblue, 0.05% Xylene Cyanole by heating at 90°C for 2 min, immediately transferred onto wet ice, and loaded on a 5% polyacrylamide gel (160 x 140 x 0.75 mm) containing 90 mM Tris-borate, pH 8.3, 4 mM EDTA for SSCP analysis as described by Orita et al. (1989). Electrophoresis was in the 200μ1 vertical electrophoresis chamber (LKB) at 40 W for 1 hr with cooling to 20°C. The same gel system was used to separate undenatured K1 PCR products. The K10 alleles were separated following denaturation as above on 5% polyacrylamide gels containing also 5.75 M urea (40 W, 1.5 hr, 20°C). Detection was by silver staining. Except for their retarded migration velocity, single strands could be easily identified as such by their more reddish color upon drying, while reassociated double strands that occurred occasionally in overloaded lanes stained black.

RESULTS AND DISCUSSION

Analysis of keratin enriched cytoskeletal residues obtained from different human epithelia by high resolution SDS gel electrophoresis has shown that certain keratin polypeptides, namely the type II keratins K1, K4, and K5 and the type I keratin K10, display inter-individual variations within the corresponding polypeptide patterns. Accordingly, we have proposed that all these keratin genes are polymorphic and express codominant alleles. (Wild and Mischke, 1986; Mischke and Wild, 1987; Mischke et al., 1990).

Because the main features of intermediate filament chain structure suggest that the N- and C-terminal domains in particular define the members of each class by their characteristic size, composition, and sequence, we analyzed these domains by means of the polymerase chain reaction, single strand conformation gel electrophoresis, and sequencing, in order to determine the molecular basis of the genetic polymorphism of these keratins (Korge et al., 1992a and 1992b; Wanner et al., 1993).

The alleles of K4 and K5 differ in their N-terminal domains

When genomic DNA from individuals homozygous for the K4a or K4b phenotype at the protein level was amplified with primers specific for the N-terminal domain, two amplification products of different sizes (177 bp and 135 bp, respectively, with primers K4-20a and K4-20b) were obtained. In heterozygous individuals the corresponding doublet of bands was detected. SSCP analysis (Figure 1a) furthermore revealed two pairs of single strands for the K4a allele that were named K4a1 and K4a2, accordingly.

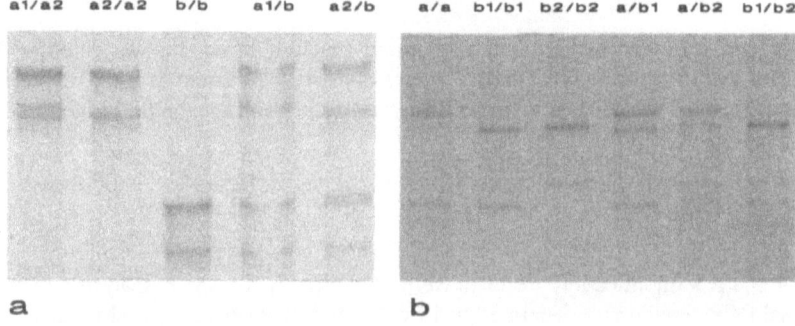

Figure 1. SSCP analysis of the PCR products from the N-terminal domains of K4 and K5. (a) Alleles of K4. Silver stained gel showing five of six possible combinations of three alleles in homozygous and heterozygous genotypes with sizes of 177 bp (K4a1, K4a2) and 135 bp (K4b). (b) Alleles of K5. Silver stained gel of the 251 bp amplicon generated with the primer pair K5-11d and K5-11b from homozygous and heterozygous genotypes. Apparently, the upper single strand of K5b1 and K5b2 as well as the lower single strand of K5a and K5b1 show the same migration velocity

Sequencing indicated that the difference between K4a1 and K4a2 is due to a nucleotide transition from T to C and between alleles K4a2 and K4b to a 42 bp deletion. The deduced amino acid sequence showed the deletion to be in frame and equivalent to the loss of 14 amino acids, i.e. 3 glycine loops in the V1 subdomain (Wanner et al., 1993). As proposed by Steinert et al. (1991), these glycine loops may be involved in the structural organization of keratin filaments in epithelial cells and thus contribute to the flexibility of the entire tissue.

The differences among the K5 alleles also reside in the N-terminal domain and a total of three pairs of single strands could be discerned on SSCP-gels (Figure 1b). Two of those belonged to the protein phenotype K5b and were therefore labelled K5b1 and K5b2. For these, direct sequencing revealed a third base C to T transition having no effect at the amino acid level of the V1 subdomain. The alleles K5a and K5b1 were distinguished by a transversion from A to G replacing a glycine with a glutamic acid in the H1 subdomain (Wanner et al., 1993).

The alleles of K1 and K10 differ in their C-terminal domains.

In contrast to K4 and K5, the allelic differences among the alleles of the K1 and the K10 genes, which are expressed in the suprabasal cells of keratinizing tissues such as the interfollicular epidermis, do not reside in the N-terminal but rather in the C-terminal domain. They present as size alleles with amplification products that directly reflect the size differences of the corresponding polypeptides (Korge et al., 1992a and 1992b). For example, the K10 alleles are defined by deletions (or insertions) encoding one or more glycine loops (Steinert et al., 1991) that are restricted to three major sites along the C-terminal V2 subdomain (Korge et al., 1992a). Similarly, the two K1 alleles differ by a 21 bp deletion, i.e., one glycine loop of seven amino acids within the carboxyl-terminal V2 subdomain (Korge et al., 1992b).

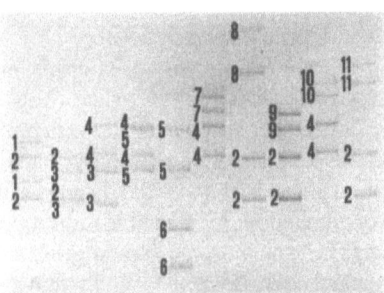

Figure 2. Single strand separation of PCR products from the polymorphic C-terminal domain of K10. Genomic DNA was amplified with primers K10-1a and K10-1b and separated on 5% polyacrylamide/5.75 M urea gels. Under these conditions the single strands showed a conformation dependent migration behavior allowing for consistent identification of all alleles after silver staining. Only heterozygous individuals are shown with the alleles labelled accordingly (Mischke, 1993).

Since the terminal domains of keratins contain many repeated motifs, we have found very short annealing and elongation times and the inclusion of glycerol and formamide sometimes helpful to reduce self priming of products due to such repeats. For example, cycle settings of 1 min at 94°C, 2 sec at 58°C, and 5 sec at 72°C were used for a „Speed-up PCR".

The size difference between the largest and the smallest K10 allele was determined to be 114 bp or 38 amino acid (Korge et al., 1992a). The number of alleles, however, was found to exceed the number of variants detected by protein analysis (Mischke and Wild, 1987) by far. Since there were also some more subtle size differences among the alleles difficult to resolve reliably on non-denaturing gels, single strands were separated on TBE/urea polyacrylamide gels. Under these conditions consistent identification of the alleles could be achieved (Figure 2). Within a Caucasian population of European origin 11 alleles of K10 were identified. Of these, alleles K10.2 and K10.3 encoded the previously described polypeptide variant K10b and alleles K10.4 and K10.5 the variant K10a (Mischke, 1993). Their appropriately combined allele frequencies corresponded well to those obtained for the polypeptide variants in a different sample (Mischke and Wild, 1987).

Allele frequencies of human polymorphic keratins

The allele frequencies for all alleles were determined in a Caucasian population of European origin (Table 2). As expected, the highly polymorphic K10 gene showed an observed heterozygosity of 74%. The resulting PIC value of 0.68 (Mischke, 1993) is almost as high as those for microsatellite DNA-polymorphisms suggesting that the K10 alleles should be highly informative markers for linkage analysis in the type I keratin gene cluster on chromosome 17q in hereditary skin diseases like epidermolytic palmoplantar keratoderma (Reis et al., 1992) or epidermolytic hyperkeratosis (Rothnagel et al., 1992).

Table 2. Human polymorphic keratins.

Keratin	Number of alleles	observed heterozygosity	PIC	Mutation
K1	2	42%	0.37	deletion
K4	3	48%	0.44	deletion/substitution
K5	3	36%	0.33	substitution
K10	11	74%	0.68	deletion

The type II keratin genes appear to be less polymorphic, displaying only two or three alleles and hence, yield lower heterozygosity indices and PIC values (Table 2). However, the genetic polymorphism of K1, K4, and K5 together with the allelic variations of the K2 and K3 genes (M. Zirra and D. Mischke, unpublished), should also improve haplotype analyses for the type II keratin gene cluster in families affected with inherited disorders of epithelial differentiation.

In general, due to the lower likelihood of recombination as a result of close physical association or direct involvement, the polymorphic keratin genes can be expected to be more informative than anonymous markers outside of the two keratin gene clusters. Accordingly, the polypeptide polymorphism of K5 has already been helpful in unravelling the case of one epidermolysis bullosa family (Lane et al., 1992) as has the K1 polymorphism in a family with epidermolytic hyperkeratosis (Compton et al., 1992; Chipev et al., 1992).

The question as to when in evolution these alleles arose remains open. However, since the K4 and K5 genes are linked on the same chromosome but show lack of linkage disequilibrium (Mischke et al., 1991), the alleles must have evolved already very early.

In contrast to the point mutations occurring in the conserved helix initiation or helix termination peptides at the beginning or end of the rod domains of K5, K10, and K14 which cause severe blistering of the skin, as in epidermolysis bullosa simplex or epidermolytic hyperkeratosis (Bonifas et al., 1991; Coulombe et al., 1991; Lane et al., 1992; Rothnagel et al., 1992; Cheng et al., 1992), none of the described alleles and their underlying sequence differences seem to be associated with an apparent impairment of epithelial development and differentiation. Similarly, the mutation in the H1 subdomain that alters the glycine of allele K5a to the glutamic acid of allele K5b appears also to be tolerated, whereas a leucine to proline mutation located only 11 amino acids further upstream in the same subdomain of the K1 gene causes defective keratin filament formation in one family with epidermolytic hyperkeratosis (Chipev et al., 1992). Therefore, the functional and physiologic consequences of allelic variations on intermediate filament formation and, consequently, the „quality" of epithelial tissues will merit further investigations.

In conclusion, genetic polymorphism of the keratin genes K1, K2, K3, K4, K5, and K10 is rendering an already complex multigene family even more complex. The identification of alterations in keratin genes leading to disease phenotypes will, however, benefit from the improved possibilities for genetic analysis in affected families.

ACKNOWLEDGEMENTS

Funding from the Deutsche Forschungsgemeinschaft (Mi 210/6-1) is gratefully acknowledged.

We thank Gabriele Wille, Irmgard Tilmans, and Maja Zirra for expert technical assistance and Drs. J.G. Compton, H.-H. Förster, P.M. Steinert, and A. Ziegler for stimulating discussions and encouragement.

REFERENCES

Barletta C, Batticane N, Ragusa RM, Leube R, Peschle C and Romano V. Subchromosomal localization of two human cytokeratin genes (KRT4 and KRT15) by in situ hybridization. Cytogenet Cell Genet 1990, 54: 148-50

Bonifas JM, Rothman AL, and Epstein EH. Epidermolysis bullosa simplex: Evidence in two families for keratin gene abnormalities. Science 1991, 254: 1202-5

Cheng J, Syder AJ, Yu Q-C, Letai A, Paller AS and Fuchs E. The genetic basis of epidermolytic hyperkeratosis: A disorder of differentiation-specific epidermal keratin genes. Cell 1992, 70: 811-9

Chipev CC, Korge BP, Markova N, Bale SJ, DiGiovanna JJ, Compton JG and Steinert PM. A leucine > proline mutation in the H1 subdomain of keratin 1 causes epidermolytic hyper-keratosis. Cell 1992, 70: 821-8

Compton JG, DiGiovanna JJ, Santucci SK, Kearns KS, Amos CI, Abangan DL, Korge BP, McBride OW, Steinert PM and Bale SJ. Linkage of epidermolytic hyperkeratosis to the type II keratin gene cluster on chromosome 12q. Nature Genetics 1992, 1: 301-5

Coulombe PA, Hutton ME, Letai A, Hebert A, Paller AS, and Fuchs E. Point mutations in human keratin 14 genes of Epidermolysis bullosa simplex patients: Genetic and functional analyses. Cell 1991, 66: 1301-11

Eckert RL and Rorke EA. The sequence of the human epidermal 58-kD (#5) type II keratin reveals an absence of 5' upstream sequence conservation between coexpressed epidermal keratins. DNA 1988, 7: 337-45

Fuchs E. Keratins as biochemical markers of epithelial differentiation. Trends Genetics 1988, 4: 277-81

Hatzfeld M and Weber K. A synthetic peptide representing the consensus sequence motif at the carboxy-terminal end of the rod domain inhibits intermediate filament assembly and disassembles preformed filaments. J Cell Biol 1990, 110: 1199-210

Herrmann H, Hofmann I and Franke WW. Identification of a nonapeptide motif in the vimentin head domain involved in intermediate filament assembly. J Mol Biol 1992, 223: 637-50

Johnson LD, Idler WW, Zhou X-M, Roop DR and Steinert PM. Structure of a gene for the human epidermal keratin of 67 000 Dalton. Proc Natl Acad Sci USA 1985, 82: 1896-1900

Korge BP, Gan S-Q, McBride OW, Mischke D, and Steinert PM. Extensive size polymorphism of
the human keratin 10 chain resides in the C-terminal V2 subdomain due to variable numbers
and sizes of glycine loops. Proc Natl Acad Sci USA 1992a, 89: 910-914

Korge BP, Compton JG, Steinert PM, and Mischke D. The two size alleles of human keratin 1 are
due to a deletion in the glycine-rich carboxyl-terminal V2 subdomain. J Invest Dermatol 1992b,
99: 697-702

Lane EB, Rugg EL, Navsaria H, Leigh IM, Heagerty AHM, Ishida-Yamamoto A, and Eady RAJ. A
mutation in the conserved helix termination peptide of keratin 5 in hereditary skin blistering.
Nature 1992, 356: 244-246

Lersch R and Fuchs E. Sequence and expression of a type II keratin, K5, in human epidermal cells.
Mol Cell Biol 1988, 8: 486-493

Lessin RS, Hübner K, Isobe M, Croce CM, and Steinert PM. Chromosomal mapping of human
keratin genes: Evidence of non-linkage. J Invest Dermatol 1988, 91: 572-578

Leube R, Bader BL, Bosch FX, Zimbelmann R, Achtstaetter T, and Franke WW. Molecular
characterization and expression of the stratification-related cytokeratins 4 and 15. J Cell Biol
1988, 106: 1249-1261

Maniatis T, Fritsch EF, and Sambrook J. Molecular Cloning: A Laboratory Manual, Cold Spring
Harbor Laboratory, New York, 1982

Mischke D and Wild G. Polymorphic keratins in human epidermis. J Invest Dermatol 1987, 88: 191-
197

Mischke D, Wille G, and Wild AG. Allele frequencies and segregation of human polymorphic
keratins K4 and K5. Am J Hum Genet 1990, 46: 548-552

Mischke, D. Frequencies of human keratin 10 alleles. Hum Molec Biol 1993, 2: 618

Moll R, Franke WW, Schiller DL, Geiger B, and Krepler R. The catalog of human cytokeratins:
patterns of expression in normal epithelia, tumors and cultured cells. Cell 1982, 31: 1-24

Orita M, Suzuki TS, and Hayashi, K. Rapid and sensitive detection of point mutations and DNA
polymorphisms using the polymerase chain reaction. Genomics 1989, 5: 874-879

Oshima RG. Intermediate filament molecular biology. Curr Opin Cell Biol 1992, 4: 110-116

Parry DAD and Steinert PM. Intermediate filament structure. Curr Opin Cell Biol 1992, 4: 94-98

Reis A, Küster W, Eckhardt R, and Sperling K. Mapping of a gene for epidermolytic palmoplantar
keratoderma to the region of the acidic keratin gene cluster at 17q12-21. Hum Genet 1992, 90:
113-116

Rieger M and Franke WW. Identification of an orthologous mammalian cytokeratin gene. J Mol Biol 1988, 204: 841-856

Rosenberg M, Fuchs E, Le Beau MM, Eddy RL, and Shows TB. Three epidermal and one simple epithelial type II keratin genes map to human chromosome 12. Cytogenet Cell Genet 1991, 57: 33-38

Rothnagel JA, Dominey AM, Dempsey LD, Longley MA, Greenhalgh DA, Gagne TA, Huber M, Frenk E, Hohl D, and Roop DR. Mutations in the rod domains of keratins 1 and 10 in epidermolytic hyperkeratosis. Science 1992, 257: 1128-1139

Steinert PM and Roop DR. Molecular and cellular biology of intermediate filaments. Annu Rev Biochem 1988, 57: 593-625

Steinert PM, Mack JW, Korge BP, Gan S-Q, Haynes SR, and Steven AC. Glycine loops in proteins: their occurrence in certain intermediate filament chains, loricrins and single-stranded RNA binding proteins. Int J Biol Macromol 1991, 13: 130-139

Sun T-T, Eichner R, Schermer A, Cooper D, Nelson WG, and Weiss RA. Classification, expression, and possible mechanisms of evolution of mammalian epithelial keratins: a unifying model. In: Levine A, Topp W, Vande Woude G, Watson JD (eds) Cancer Cells 1: The transformed phenotype. Cold Spring Harbor Laboratory, New York, 1984, pp 169-176

Wanner R, Förster H-H, Tilmans I, and Mischke D. Allelic variations of human keratins K4 and K5 provide polymorphic markers within the type II keratin gene cluster on chromosome 12. J Invest Dermatol 1993, 100: 735-741

Wild G-A and Mischke D. Variation and frequency of cytokeratin polypeptide patterns in human squamous non-keratinizing epithelium. Exp Cell Res 1986, 162: 114-126

Zhou X-M, Idler WW, Steven AC, Roop DR, and Steinert PM. The sequence and structure of human keratin 10: organization and possible structures of end sequences. J Biol Chem 1988, 263: 15584-15589

PRODUCTION OF ANTIBODIES OF MONOCLONAL SPECIFICITY WITHOUT THE USE OF HYBRIDOMA CELL LINES

T.R. Gingeras[1], P. Koutz[1], P.-J. Linton[2], D.J. Decker[2], N.R. Klinman[2] and C. Stillman[1]

[1]Baxter Diagnostics Inc., Life Sciences Research Laboratory, San Diego, USA; [2]Scripps Research Institute, La Jolla, USA

INTRODUCTION

Reproducible immunodiagnostic tests require antibodies of standardized specificity and affinity. A significant step toward this goal has been achieved with the development of the methodology of growing clonal populations of cells secreting antibodies with a defined specificity (Köhler and Milstein, 1975). In this technique an antibody-secreting plasma cell, isolated from an immunized animal, is fused with an immortal myeloma cell. The products of this fusion are called hybridoma cells and are the source of monoclonal antibodies which currently provide the most reproducible and well characterized antibodies for immunodiagnostics.

Development of hybridoma cell lines producing monoclonal antibodies of a desired specificity requires multiple experimental steps which may extend over a significant time period. The steps involved in production of a monoclonal antibody include immunization of the animal, screening of the hybridoma, and antibody production (Harlow and Lowe, 1988). Any of these steps may pose problematic challenges in which costs increase in proportion to the amount of time and effort needed. Additional difficulties are encountered frequently even after a monoclonal hybridoma cell line is identified and clonally expanded. An expanded population of hybridoma cells can lose its ability to produce a specific monoclonal antibody after prolonged growth in culture. This propensity to lose antibody expression is also observed in hybridoma cell lines that have been frozen and stored after clonal selection. Thus, the stability of valuable hybridoma cell lines is not assured.

Methods in DNA Amplification, Edited by
A. Rolfs *et al.*, Plenum Press, New York, 1994

One approach to address this stability problem has been to clone the light and heavy chain genes for a specific monoclonal antibody from an hybridoma mRNA pool. In so doing, a stable supply of a specific monoclonal antibody could be ensured by virtue of the stability of expression in a *Escherichia coli* host cells. Orlandi et al., (1989) and Sastry et al., (1989) were the first to employ a reverse transcriptase (RT)/polymerase chain reaction (PCR) methodology (Kawasaki, 1990) to clone and express the variable domains of light and heavy chains of monoclonal antibodies in *E. coli*.

However, several factors suggest that this cloning approach may not always prove useful or productive. First, a hybridoma cell line must be identified and established to serve as a target for the PCR amplification and cloning steps. As indicated above, this step of hybridoma cell line identification may prove to be problematic. Second, reverse transcriptase (RT)/PCR methodology used to clone light and heavy chain mRNA from a hybridoma cell line producing a monoclonal antibody of interest has at times proven unsuccessful. This failure to clone desired immunoglobulin sequences has been ascribed to the need to thermocycle during the PCR protocol. The denaturation steps used in RT/PCR can result in the amplification of both DNA and RNA sequences present in the original target nucleic acid pool. This amplification of undesirable DNA sequences increases the complexity of the amplification-generated cDNA library. In turn, this significantly increases the likelihood of missing clones containing the productive immunoglobulin rearrangements.

SPLEEN FRAGMENT CULTURE AND SELF-SUSTAINED SEQUENCE REPLICATION (3SR) TECHNIQUES: A METHOD OF PRODUCING MONOCLONAL ANTIBODIES WITHOUT THE USE OF HYBRIDOMA CELL LINES

The problems associated with the use of hybridoma cell lines for monoclonal antibody production have been addressed by devising an alternative approach to produce such antibodies. This approach employs a combination of two techniques: murine spleen fragment culture and the self sustained sequence replication (3SR) method of RNA-specific amplification.

Spleen Fragment Culture

Secondary (2° or memory) B cells differ in several ways from primary (1°) B cells in their response to the same antigen by: a) their expression of surface isotypes other than IgM and IgD (Black et al., 1977, Teale et al., 1981); b) their low expression of cell surface antigen recognized by the monoclonal antibody J11D (Bruce et al., 1981); c) their capacity to recirculate throughout the lymphatic system (Strober, 1972); d) their requirements for stimulation and burst size (Klinman, 1972); and, perhaps most significantly; e) their repertoire of expressed variable regions (arising from somatic mutation of this region) that dominate a response (Allen et al., 1987).

Murine spleen fragment cultures (Klinman, 1974; Riley and Klinman, 1985) can be used to generate *in vitro* monoclonal AFC responses from either 1° or 2° B cells (Figure 1). Briefly, limiting numbers of B cells obtained from either naive or previously immunized mice were transferred intravenously into lethally irradiated (1300R) MHC syngeneic mice. The transferred B cells can be further fractionated into cells which possess low levels of the surface marker recognized by the J11D monoclonal antibody (Bruce, et al. 1981)   (Figure 1). In this study, J11D^{lo} precursor cells were not fractionated prior to transfer into the irradiated mouse. The transferred B cells colonized the mouse spleen, and within 24 hours the spleen was removed and dissected into 1 mm cubic fragments. Each fragment was cultured in individual wells of a microtiter dish in the presence of antigen for 2-3 days. Culture fluids were screened for antibodies specific for the stimulating antigen 5-7 days after the removal of antigen from the culture. Cultures which were antibody positive were subjected to phenol-chloroform extraction of total nucleic acids (DNA and RNA). The total nucleic acid was used for the 3SR amplification and subsequent cloning of light and heavy immunoglobulin c-DNAs.

Specific Amplification of Murine Spleen Fragment RNA by 3SR Amplification

The *in vitro* target amplification of RNA under isothermal conditions was first described by Guatelli, et al. (1990), Gingeras, et al. (1990), and later modified by Fahy, et al. (1991). This amplification method has been employed to clone mRNA copies of the light and heavy chain immunoglobulin genes from cultured murine spleen fragments that express antibodies specific for estradiol. Because the nucleotide sequences of the anti-estradiol heavy and light chain mRNAs were unknown generic primers were developed for the 3SR amplification of these targets. Nucleotide sequences of the generic light and heavy chain primers used in the 3SR reaction as well as the position of the oligonucleotide probes used in the detection of the amplified products are shown in Figure 2A and Figure 2B, respectively.

Total nucleic acid extracted from spleen fragment cultures expressing anti-estradiol antibodies was first amplified by 3SR, and the RNA products were then converted to DNA for subsequent cloning (see legend to Figure 3). The cDNA products of the amplification of light and heavy chain mRNAs were characterized by Southern hybridization. Bands corresponding to the expected the expected sizes of 400 and 430 base pairs for the variable heavy chain and leader heavy chain amplification products, respectively, are visible. The longer 465 base pair amplification product observed from the 3SR reaction using the leader heavy chain primers is
    the result of moving the 5' end primer approximately 50 base pairs upstream from the variable region primers. The variable light chain amplification product was 340 base pairs (data not shown).

Analysis of Light and Heavy Chain Sequences Derived from Clones Isolated from a Single Spleen Fragment Culture

Culture media from one spleen fragment culture, D_1A_6 showed a modest affinity constant of approximately $10^{-7}M$ during initial screening steps as assessed by inhibition of antibody binding to antigen-coated plates in the presence of varying concentrations of soluble antigen.

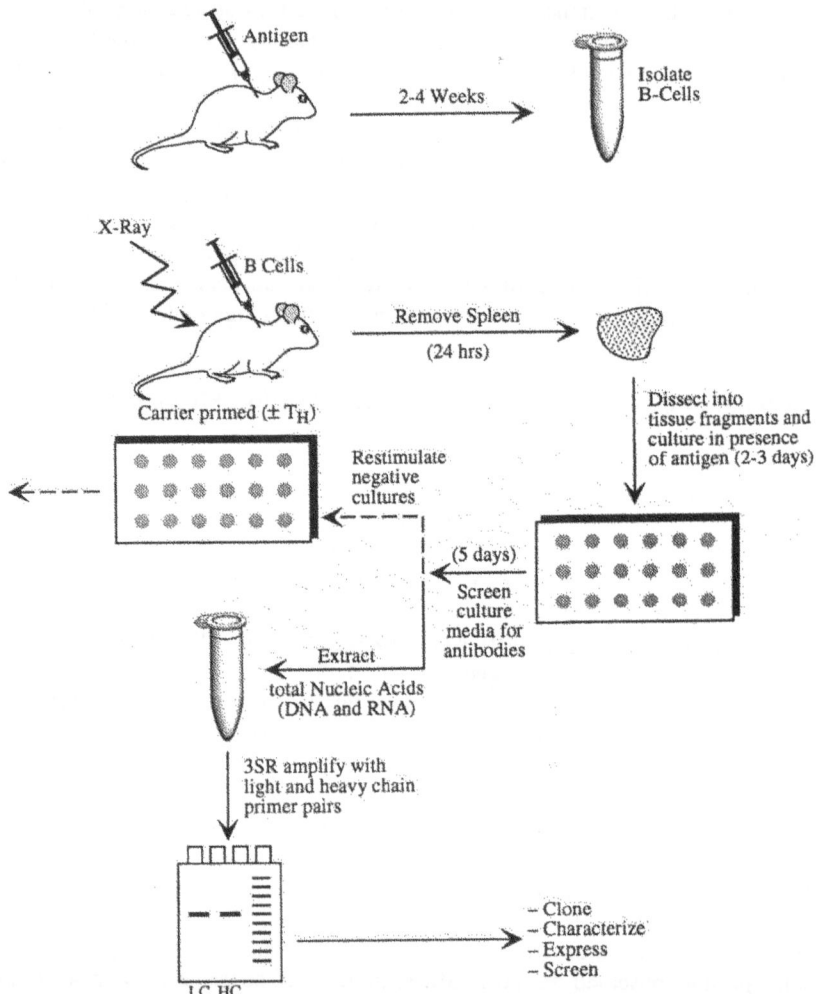

Figure 1. Scheme of 3SR/Murine Spleen Fragment Production of Monoclonal Antibodies. Steps in protocol explained in text. Condition for 3SR reaction and for cDNA synthesis used in cloning described in the legend to Figure 3.

The total nucleic acids from the cells in this spleen fragment was extracted and amplified with the light and heavy chain generic primers previously described (Figure 2A and 2B). The amplification products were cloned into pBR322, and the nucleotide sequence of 33 light and 36 heavy chain clones were determined. Interestingly, based on the nucleotide sequences of the variable regions and the V, D, J hypervariable regions, four groups of heavy chain genes could be identified. Of these four groups, only one group ($^{10}/_{33}$ clones) was homogeneous in nucleotide sequences comparing the nucleotide sequences of the complementarity determining regions (CDRs), as well as the V, D, J junction regions. Additionally, the putative amino acid sequence derived from the nucleotide sequences indicated that all members of this homogeneous group are potentially functional clones (i.e., without the presence of termination codons in the variable or hypervariable regions). Of the other three heavy chain groups, two groups contained only non-functional clones and one group, while predicted to be from a productive rearrangement, was heterogeneous in the nucleotide sequences of the CDR and hypervariable regions. The homogeneous heavy chain group of clones contained JH2 sequences. Comparison of nucleotide sequences to other mouse heavy chain variable region sequences (V_H) (Kabat, et al. 1992), indicated a 79% homology to the V_H S107 family.

The light chain clones were analyzed in a manner similar to heavy chain clones. Based on the nucleotide sequences of the V, J and hypervariable CDR regions, three groups of light chains were identified. The three groups of light chains used predominantly J kappa 5 sequences. The putative amino acid sequence of the three groups revealed only two sequence groupings, of which only one group ($^{20}/_{36}$ clones) was composed of potentially functional clones. The remainder of the clones were nonproductive. Nucleotide homology to V kappa H9 (92%) and H3 (91%) genes was noted for clones of the productive light chain group.

This analysis of the heavy and light chain clones from a single spleen fragment culture suggests two conclusions. First, the light and heavy chain sequences which were the most abundantly represented in the clones analyzed and which were predicted to be derived from productive rearrangements, most likely represent the original paired light and heavy chains detected in the original D_1A_6 culture media. Second, the generic 3SR light and heavy chain primers were used successfully to amplify mRNAs from multiple B cells expressing heavy and light chain genes. The origin of the four heavy chain and three light chain groups cannot be unambiguously determined. One possibility is that more than one B cell colonized the spleen fragment, D_1A_6; however, only one of the potential colonizing B cells generated AFCs which expressed the antibody detected during the initial screening process. The second possibility is that there was only one precursor B cell that colonized in the D_1A_6 spleen fragment, but the radiation treatment used on the recipient mouse did not completely ablate the host mouse immune system. Consequently, the observed multiple groups of heavy and light chain clones may reflect low level immunoglobulin mRNA expression from the irradiated host mouse B cells. Heavy and light chain clones from other cultured spleen fragments are being analyzed to determine if multiple groups of heavy and light chains are also observable in these cultures.

ADVANTAGES OF USING THE SPLEEN FRAGMENT/3SR METHODOLOGY IN THE CLONING OF IMMUNOGLOBULINS

Several reviews have described the cloning and the engineering of antibodies beginning with chromosomally encoded nucleotide sequences (Hoogenboom, et al., 1992, Larrick, et al., 1992, Clackson, et al., 1991). The approaches described in these reviews begin with either antigen-stimulated B cells isolated from the peripheral circulatory system or with antibody expressing hybridoma cell lines. This initial cell isolation step is routinely followed by sequence specific amplification using RT/PCR followed by the cloning of the amplification products as Fab or single chain antibodies. When peripheral circulating B cells are used as targets to obtain clones of the antibody of interest, the light and heavy chain clones are identified by screening with the original stimulating antigen using a phage display system (Clackson et al., 1991; Kang, 1991; Huse et al., 1989). This step is necessary in order to

A.   **Generic Heavy Chain Primers and Probes**

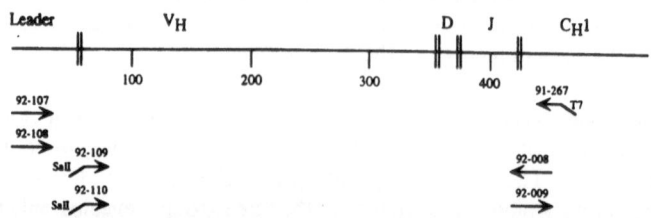

5' End Leader Region Primers

92-107 5'-ATG (GA)A(GC) TT(GC) (TG)GG (TC)T(AC) A(AG)C T(GT)G (GA)TT-3'
92-108 5'-ATG (GA)AA TG(GC) A(GC)C TGG GT(CT) (TA)T(TC) CTC T-3'

5' End Variable Region Primers

92-109 5'-(GC)AG GT(CG) (AC)A(AG) CTG CAG (CG)AG TCT-3'
92-110 5'-(CG)AG GTG (CA)AG CTC (CG)(AT)(AG) (CG)A(AG) (CT)C(CG) GGG-3'

3' End Constant Region Primers

91-267 5'-AATTTAATACGACTCACTATAGGGAAGTGGATAGACAGATGGGGGTG-3'

Sense Strand Probe

92-009 5'-TCACCGTCTCCTCAGCCAAAACGA-3'

B   **Generic Light Chain Primers and Probes**

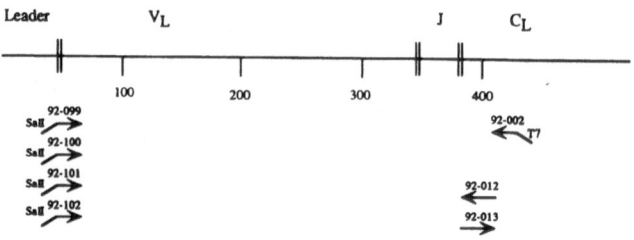

5' End Variable Region Primers

92-099 5'-GA(TC) AT(TC) GT(GC) CT(CG) AC(ACG) CA(AG) TC$\underline{A}$ CCA-3'
92-100 5'- . . . . $\underline{T}$ . . . -3'
92-101 5'- . . . . $\underline{C}$ . . . -3'
92-102 5'- . . . . $\underline{G}$ . . . -3'

3' End Constant Region Primers

92-002 5'-AATTTAATACGACTCACTATAGGGAGCTCACTGGATGGTGGGAAGATG-3'

Sense Strand Probe

92-013 5'-GCTGATGCTGCACCAACTGTATCC-3'

Figure 2. Detailed explanation of the figure is given on the next page

Figure 2. Illustration of the heavy (**A**) and light (**B**) immunoglobulin genes showing the position of 3SR, PCR, and detection oligonucleotides. **A.** The leader heavy chain variable region was amplified by 3SR with primers 92-(107,108) and 91-267. Generic primers 92-(107,108) hybridize to the 5' end of the heavy chain leader region, and the specific primer 91-267 hybridizes to the 3' end in the heavy chain constant region. Primer 91-267 contains a T7 promoter sequence at its 5' end. Nucleotides listed inside the ( ) indicate that any of those nucleotides can be present at this position without the primer. An aliquot of the 3SR RNA amplification product was subjected to an RT/PCR amplification using primers 92-(107,108) and a primer identical to 91-267 which does not contain the T7 promoter sequences, in order to convert the 3SR RNA products into clonable cDNA. Similarly, an aliquot of the 3SR product was subjected to a nested RT/PCR amplification with primers 92-(109,110) and primer identical to 91-267 for the amplification of the heavy chain variable region. Detection of the heavy chain gene by Southern analyses was performed on the PCR reactions with oligonucleotide 92-009, which hybridizes to the 3' region of the heavy chain variable region. **B.** The light chain variable region was amplified by 3SR with primers 92-(099-102) and 92-002. Generic primers 92-(099-102) hybridize to the 5' end of the light chain variable region, and the specific primer 92-002 hybridizes to the 3' end in the light chain constant region. Primer 92-002 contains a T7 promoter binding site at its 5' end. An aliquot of the 3SR RNA amplification product was also subjected to an RT/PCR amplification to generate clonable cDNA copies using primers 92-(099-102) and 92-012, which hybridize to the 5' end of the light chain variable region and the 3' end in the light chain constant region, respectively. Detection of the light chain gene by Southern analyses was performed on the PCR reactions with oligonucleotide 92-013 which hybridizes to the light chain constant region.

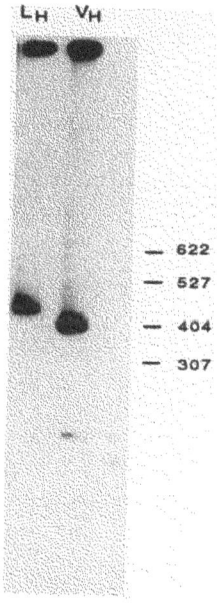

Figure 3. Southern blot analysis of the anti-estradiol (anti-E2) heavy immunoglobulin chain 3SR/PCR products. The filter was probed with sense oligonucleotide 92-009 (Figure 2). An *Msp*I digest of pBR322 DNA was used for molecular weight markers (lane 3). The 3SR reaction was carried out as described by Fahy, et al (1991) with the following modifications. Reverse transcriptase (RT) T7 RNA polymerase and *E. coli* RNaseH were added at 15, 50 and 1 units, respectively in a total reaction volume of 50 ml. The 3SR RNA products were converted to clonable double-stranded cDNA using RT/PCR protocol supplied by Perkin Elmer. The thermal cycle parameters used were 30 cycles at 94°C (1 min.), 55°C (1 min.) and 72°C (2 min.) followed by 10 min. at 72°C and 4°C for 1 min. The anti-E2 leader heavy chain variable region was amplified from total nucleic acids from a secondary murine spleen fragment by 3SR with primers 92-(107,108)/91-267, then subjected to RT/PCR amplification with primers 92-(107/108)/primer analogous to 91-267, and yielded an amplified fragment of the expected size (approximately 430 bp) for the leader heavy chain variable region (lane 1). The 3SR product from the amplification of the leader heavy chain variable region was subjected to a nested RT/PCR amplification with primers 92-(109,110)/primer analogous to 91-267, and yielded an amplified fragment of the expected size (approximately 400 bp) for the heavy chain variable region (lane 2).

identified by screening with the original stimulating antigen using a phage display system (Clackson, et al., 1991, Kang, 1991, Huse, et al., 1989). This step is necessary in order to identify the original light and heavy chain pairings that were elicited within the B cells at the time of sensitization. The murine spleen fragment/3SR methodology of cloning antibody genes provides an alternate method to identify such original light and heavy chain pairings. In addition several other advantages accompany the use of this methodology including the ability to select for: 1) specific isotypes, 2) antibodies with binding affinities of desired levels, and 3) defined nature using competitive analog screening. All of these selections can be performed through the analysis of the antibodies produced by the fragment cultures before confronting the cloning, characterization, and recombinant expression challenges that follow identification of AFC producing antibodies of monoclonal specificity.Finally, the maturation of antibody producing B cells is a subject studied with considerable interest. Mechanistically both somatic mutation and affinity based antigen selection are poorly understood. However, the ability to analyze these processes during different stages of B cell maturation is possible using murine spleen fragment culture method. In combination with the 3SR amplification methodology, the processes of somatic mutation and B cell development is being studied after each exposure to a stimulating antigen (Decker, et al., 1993).

ACKNOWLEDGMENTS

The authors thank G. Davis and D. McElligott for critical reading of the manuscript and C. Vavrunek for her careful preparation of this manuscript. Partial support for this work was provided to N.R. Klinman by NIH in AI15797 and by Baxter Diagnostics 92-02.

REFERENCES

Allen D, Cumano A, Dildrop R, Kocks C, Rajewsky K, Rajewsky N, Roes J, Sabilitzky F, and Siekevitz M. Genetic requirements and functional consequences of somatic hypermutation during B cell development. Immunol Rev 1987, 96:5-22.

Berek C, Griffith GM, Milstein C. Molecular events during maturation of the immune response to oxazolone. Nature 1985, 316:412-18.

Black SJ, Van der Loo W, Loken R, and Herzenberg LA. Expression of IgD by murine lymphocytes. Loss of surface IgD indicates maturation of memory B cells. J Exp Med 1977, 147:984-96.

Bruce J, Symington FW, McKearn TJ, and Sprent J A monoclonal antibody discriminating between subsets of T and B cells. J Immunol 1981, 127:2496-501.

Clackson T, Hoogenboom HR, Griffiths AD, and Winter G. Making antibody fragments using phage display libraries. Nature 1991, 352:624-28.

Decker DJ, Linton P-J, Jacobs S, Biery M, Gingeras TR, and Klinman NR. Defining subsets of naive and memory B cells based on their ability to somatically mutate *in vitro*. 1993, Submitted for publication.

Fahy E, Kwoh DY, Gingeras TR. 3SR: An isothermal transcription-based amplification alternative to PCR. PCR Methods and Appl. 1991, 1:25-32.

Gingeras TR, Whitfield KM, and Kwoh DY. Unique features of the self-sustained sequence replication (3SR) reaction in the *in vitro* amplification of nucleic acids. Ann Biol Clin 1990, 48:498-501.

Guatelli JC, Whitfield KM, Kwoh DY, Barringer KJ, Richman DD, and Gingeras TR. Isothermal *in vitro* amplification of nucleic acids by multi-enzyme reaction modeled after retroviral replication. Proc Natl Acad Sci USA 1990, 81:1874-78.

Harlow E and Lowe D. In: Antibodies. Cold Spring Harbor Press, Cold Spring Harbor, N.Y. 1988, p. 149.

Hoogenboom HR, Marks JD, Griffiths AD, and Winter G. Building antibodies from their genes. Immunol Rev 1992, 130:41-68.

Huse WD, Sastry L, Iverson SA, Kang AS, Alting MM, Burton DR, Benkovic SJ, and Lerner RA. Generation of a large combinatorial library of the immunoglobulin repertoire in phage lambda. Science 1989, 246:1275.

Kabat EA, Wu TT, Perry HM, Gottesman KS, and Foella C. In: Sequences of Proteins of Immunological Interest (5th Edition) Volume I. U.S. Department of Health and Human Services, NIH Public Health Service, Publication No. 91-3242, 1991.

Kang A S, Barbas CF, Janda KD, Benkovic SJ, and Lerner RA. Linkage of recognition and replication functions by assembling combinatorial antibody Fab libraries along phage surfaces. Proc Natl Acad Sci USA 1991, 88:4363.

Kawasaki ES. Amplification of RNA. In: PCR Protocols: A guide to methods and applications. (Eds: M.A. Innis, D. Gelfand, J.J. Sninsky, and T.J. White) 1990, p. 21-27.

Klinman N R. The mechanism of antigen stimulation of primary and secondary clonal precursor cells. J Exp Med 1972, 136:241-60.

Klinman N R, Press J L, Pickard AR, Woodland RT, and Dewey AF. (1974). Biography of the B cell. In: The Immune System (Eds. E. Sercarz, A. Williamson, and C. F. Fox). New York, Academic Press 1991, p. 357-65.

Kohler G, and Milstein C. Continuous cultures of fused cells secreting antibody of predefined specificity. Nature 1975, 256:495.

Larrick JW, Wallace EF, Coloma MJ, Bruderer U, Lang AB, and Fry KE. Therapeutic human antibodies derived from PCR amplification of B cell variable regions. Immunol Rev 1992, 130:69-85.

Linton PJ, Decker D, Klinman N. Primary antibody forming cells and secondary B cells are generated from separate precursor cell subpopulations. Cell 1989, 59:1049-59.

Orlandi R, Gussow DH, Jones PT and Winter G. Cloning immunoglobulin variable domains for expression by the polymerase chain reaction. Proc Natl Aca Sci USA 1989, 86:3833.

Riley RK, and Klinman N. Differences in antibody repertories for (4-hydroxy-3-Nitrophenyl) acetyl (NP) in splenic versus immature bone marrow precursor cells. J Immunol 1985, 135:3050-55.

Sastry L, Alting-Mees M, Huse WD, Short JM, Sorge JA, Hay BN, Janda KD, BenKovic SJ, and , Lerner RA. Cloning of the immunological repertoire in *E. coli* for generation of monoclonal catalytic antibodies: Construction of a heavy chain variable region-specific cDNA library. Proc Natl Acad Sci USA 1989, 86:5728-32.

Strober S. Initiation of antibody responses by different classes of lymphocytes. Fundamental changes in the physiological characteristics of virgin thymus-independent ("B") lymphocytes and "B" memory cells. J Exp Med 1972, 136:851-71.

Teale JM, Lafrenz D, Klinman NR, and Strober S. Immunoglobulin class commitment exhibited by B lymphocytes separated according to surface isotype. J Immunol 1981, 126:1952-57.

USE OF CHELEX 100™ IN THE EXTRACTION OF VIRUSES FROM DIVERSE CELL-FREE CLINICAL SAMPLES FOR PCR

A. Ochert[1], M. J. Slomka[2], J. Ellis[2] and C.G. Teo[2]

[1]King's College School of Medicine and Dentistry, London, UK, [2]Central Public Health Laboratory, London, UK

INTRODUCTION

Extraction of nucleic acids from samples for PCR is often time-consuming and may involve a large number of pipetting and transferring steps, with the addition of a number of reagents. This may cause loss of valuable target material and cross-contamination, and increase the opportunity for errors. It may also reduce the number of samples that might conveniently be handled at one time.

In the standard PCR, where only a positive or negative result is required, as in a diagnostic laboratory where there may be a large number of samples to test, it is desirable to employ a rapid simple method of nucleic acid extraction. When a semi-quantitative or quantitative PCR is required, a nucleic acid extraction procedure that is reproducible and easy to maintain is vital. This is necessary because quantitative PCR relates to the original target-load in a sample, and any procedure that reduces the efficiency of recovery of the target would yield inaccurate data. Some workers have described the use of single-step extraction methods. Techniques commonly used are: heat treatment, sonication, microwaving, freeze-thawing and alkali-denaturation (van den Brule et al., 1990, Buffone 1992, Picard et al 1992, Sen et al 1992). In these methods PCR inhibitors present in the sample may not be totally removed. Possible PCR inhibitors of note are: haemoglobin and its degradation products, sputum, saliva, plant polysaccharides, amniotic fluid, serum, cerebrospinal fluid (CSF), urine and anticoagulants (Demeke et al 1992, Grimprel et al 1991, Holodniy et al 1991, Khan et al 1991, Ruano et al 1992, Soini et al 1992). Although the method we describe also includes 'whole' sample being added to the PCR, the ion-exchange resin, Chelex 100™ appears to remove the effect of inhibitors, (Singer-Sam 1989). A mechanism for this has been suggested (Walsh et al 1991). In this paper, we describe the use of Chelex 100™ in the simple and cost-effective extraction of viruses from cell-free clinical samples.

Methods in DNA Amplification, Edited by
A. Rolfs *et al.*, Plenum Press, New York, 1994

METHODS

We have examined a number of cell-free clinical and laboratory specimens which are regularly assayed by PCR for the detection of viruses. These include urine, serum, tissue culture and virus transport medium, CSF, egg fluid and saliva.

This method requires only basic laboratory equipment and materials. These include sterile ultrapure water, Chelex 100™ (200-400 mesh, sodium form, Bio-Rad Laboratories) water bath, heating block, microcentrifuge and vortexer. The following protocol describes the standard Chelex 100™ extraction method that has been applied in all our assays.

A. Chelex 100™ extraction protocol

[1. Prepare a 25% w/v stock of Chelex 100™ in sterile ultrapure water. Aliquot into about 1ml volumes and store at room temperature.]

2. To a 500 µl microcentrifuge tube, add 20 µl of resuspended 25% Chelex 100™, and prepare as many tubes as are required. Keep lids closed.

3. Add 100 µl of sample to the 20 µl stock Chelex 100™. This gives a final concentration of 4-5% of the resin.

4. Vortex briefly.

5. Incubate samples at 56°C in a water bath for 30 minutes.

6. Remove samples from the water bath and vortex briefly.

7. Incubate the tubes at 95°C in a hot block for 3 minutes, ensuring that the tube lids are firmly shut.

8. Carefully remove the samples from the hot block and vortex briefly.

9. Spin the tubes in a microcentrifuge for 3 minutes.

10. Remove between 15-30 µl of the supernatant for a 50 µl amplification.

Notes: i) In cases where inhibition of PCR is suspected, re-extract the supernatant with fresh Chelex 100™ up to 3 or 4 times if necessary. Alternatively, use an equal volume of 25% Chelex 100™ to sample and extract once.ii) 100 µl sample volumes are convenient, but can be linearly scaled-down in the protocol if the available sample is too small. iii) Centrifugation of some samples prior to extraction may be useful in the removal of debris which may be inhibitory to the PCR.

B. Procedures

Several extraction methods were compared for specimens containing virus or nucleic acid:

EBV

Two $^{1}/_{10}$ dilution series of plasmid pBR322 containing EBV BamHI K fragment were prepared in water. One series was placed directly into the PCR and the other was Chelex 100™ extracted. EBV was spiked into tissue culture fluid and whole mouth fluid from 12 healthy donors (0.2µm filtered). Ten-fold dilution series of these specimens in their respective fluids were subject to i) Chelex 100™ and ii) guanidinium thiocyanate (GuSCN) (Boom et al 1991) extractions.

HSV

HSV-1 and -2 (strains SC16 and 186 respectively) were separately spiked and diluted ten-fold with HSV-negative CSF (2pfu μl^{-1} to 2×10^{-5}pfu μl^{-1} CSF). Both series were boiled for 5 minutes and divided into 30μl volumes for extraction by i) Chelex 100™, ii) proteinase K digestion (0.25μg μl^{-1} at 56°C for 30 minutes) followed by phenol/chloroform extraction and ethanol precipitation, iii) no additional treatment (boiling only) and iv) phenol/chloroform followed by isoamyl alcohol/chloroform extraction.

Influenza A

A ten-fold dilution series in egg fluid was extracted by i) Chelex 100™ and ii) GuSCN methods.

All processed samples were then amplified by PCR which had been optimised for the relevant target region.

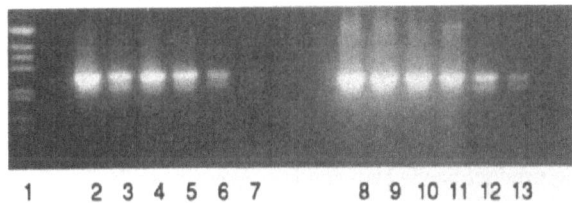

Figure 1. Dilution series of the Chelex 100™ extraction of the plasmid pBR322 containing the EBV BamHI K fragment. Lanes 2 to 7 demonstrate results after dilution in water, lanes 8 to 13 Chelex 100™ extraction prior to PCR

RESULTS AND DISCUSSION

Figure 1 shows that Chelex 100™ extraction of the plasmid (pBR322 EBV BamHI K) diution series does not cause loss of target material. Tracks 2 to 7 show the PCR products of the dilution series prepared in water and placed directly into the PCR. Tracks 8 to 13 show PCR products of the same series first subject to the Chelex 100™ extraction prior to PCR. It

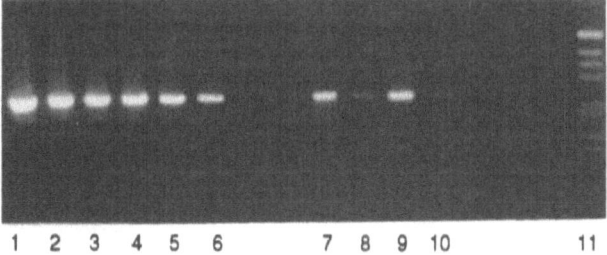

Figure 2. Comparison of the dilution series for EBV in tissue culture medium extracted by Chelex 100™ (lane 1-6) and GuSCN (lane 7-10)

appears that in some cases use of Chelex 100™ may enhance band intensity, possibly through removal of trace inhibitors or the addition into the PCR of a component of the Chelex 100™ itself. Figure 2 compares the amplification of a dilution series of EBV in tissue culture medium extracted by Chelex 100™ (tracks 1-6) and GuSCN (tracks 7-10, corresponding to 1-4). The last two dilutions extracted by GuSCN yielded no visible bands. The aliquot extracted for PCR to produce track 8 contained ten times the DNA concentration as that for track 9 but variability in the extraction has caused a greater loss of target in the sample which originally contained more DNA. This is frequently observed in the amplification of samples extracted by GuSCN. Where a small amount of target is present in a sample, any target loss will affect the outcome of a PCR more significantly. The Chelex 100™ extraction protocol is performed in a single tube so negligible loss occurs.

Figure 3 compares various extractions of HSV-2 (a) and -1 (b) in CSF, 20μl aliquots of each ten-fold dilution being amplified. Tracks 2 to 7 in both panels show specimens extracted by Chelex 100™, tracks 8 to 13 by proteinase K, phenol/chloroform and ethanol precipitiation, tracks 14 to 19 by boiling only, and 20 to 25 by phenol/chloroform only. It is clear that in this set of PCRs, phenol/chloroform extraction alone is detrimental to the production of positive PCR results even in the presence of 20pfu of virus. This may be through co-purification of PCR inhibitors, loss of target into the organic phase or interface of the phenol/chloroform mix, or most likely by traces of organic substances in the extract causing inactivation of the polymerase enzyme. Boiling alone yielded visible bands in tracks 16a and 15b giving a sensitivity line of 2pfu. Chelex 100™ extraction following boiling has further enhanced band intensity with single product bands showing clearly in tracks 2a, 3a, 4a as well as 2b and 3b. The proteinase K, phenol/chloroform, ethanol precipitation method appears to have produced results comparable to those extracted by Chelex 100™, but only in the HSV-2 PCR. In addition there are also extra non-amplicon associated bands visible which become more intense as the concentration of target decreases. How the proteinase K extraction method causes extra bands to be produced is not understood. The presence of divalent ions or some other constituent in the CSF may be responsible for this effect, which is cleared by the Chelex 100™ extraction. The proteinase K method is lengthy and includes a number of transferring

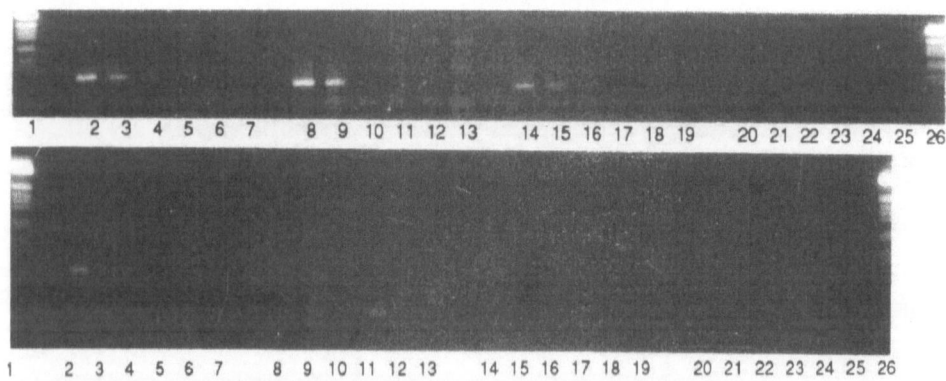

Figure 3. Comparison of various extractions of HSV-2 (a) and HSV-1 (b) in CSF. 20μl aliquots of each ten-fold dilutoin was amplified. Lane 2 to 7 in both panels demonstrate specimens extracted by Chelex 100™, lane 8 to 13 extraction by proteinase K and phenol/chloroform, lane 14 to 19 boiling only and lane 20 to 25 phenol/chloroform only.

steps as well as the addition of several reagents. Where clinical CSFs are to be examined by PCR for HSV, it is essential to exclude contamination from the extraction procedure. The importance of reproducible nucleic acid extraction procedures for diagnosis of herpes simplex encephalitis by PCR has recently been highlighted (Klapperet al 1993). The Chelex 100™ protocol appears to be reliable for the extraction of HSV DNA from CSF.

Figure 4 illustrates the effect on PCR of the presence of inhibitors. Tracks 2 and 17 are positive controls for EBV in ultrapure water, tracks 3 and 16 are blank. Tracks 4 to 15 are from filterd EBV-spiked whole mouth fluid samples extracted by Chelex 100™, and tracks 18 to 29 are the respective samples extracted by GuSCN. Apart from target loss, another major problem experienced in extraction of nucleic acids from saliva samples is the co-purification of inhibitors. This can be circumvented by Chelex 100™ extraction. The nature of the inhibitor(s) is not clear but the presence of heparin-like molecules is a possibility in this case. PCR inhibitors may be removed by repeated extraction with Chelex 100™ or by single extraction with equal volumes of sample to 25% Chelex 100™. It is important to ensure that none of the

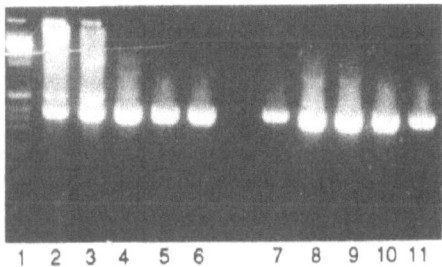

Figure 4. Effect of inhibitors on the efficacy of the PCR. Lane 2 and 17 are positive controls for EBV, lane 3 and 16 are blank. Lane 4 to 15 are from filtered EBV-spiked whole mouth fluid samples extracted by Chelex 100™, lane 18 to 29 are corresponding samples extractced by GuSCN.

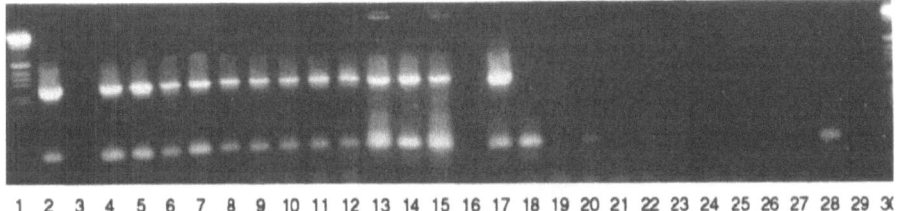

Figure 5. $1/_{10}$ dilution series of Influenza A; extraction from egg fluid by GuSCN (lane 2-6) and Chelex 100™ (lane 7-11)

Chelex 100™ is placed into the PCR tube.

Influenza A has been successfully detected in egg fluid (Fig. 5), showing that the method can also be applied to RNA viruses. Starting from neat, a $1/_{10}$ dilution series was extracted by GuSCN (tracks 2-6) and Chelex 100™. The cut-off is the same for both but the Chelex 100™ has reduced background bands whilst the product band intensity has not been affected.

CONCLUSION

This set of data has shown how a standardised, simple and relatively rapid nucleic acid extraction procedure can assist in producing reproducible, reliable PCR results. Minimal reagent preparation is required which means a reduced source of contamination. This saves time and expense in the laboratory, and unlike the GuSCN method, does not evolve hazardous waste requiring special disposal. The extraction method itself requires little 'hands-on' work, so the number of samples that can be tested at any time is not restricted. Background bands are also eliminated from PCR by following the protocol described.

This extraction method may prove a useful alternative to the proteinase K and phenol/chloroform extraction methods for cell-free material, and in cases where samples may contain inhibitors, Chelex 100™ appears to be useful in their removal.

REFERENCES

Boom R, Sol CJ, Salimans MM, Jansen CL, Wertheim-van-Dillen PM, van der Noordaa J. Rapid and simple method for purification of nucleic acids. J Clin Microbiol 1990, 28: 495-503

van den Brule AJ, Meijer CJ, Bakels V, Kenemans P, Walboomers JM. Rapid detection of human papillomavirus in cervical scrapes by combined general primer-mediated and type-specific polymerase chain reaction. J Clin Microbiol 1990, 28: 2739-2743

Buffone GJ. Improved amplification of cytomegalovirus DNA from urine after purification of DNA with glass beads. Clin Chem 1992, 38: 2360

Cheyrou A, Guyomarc`h C, Jasserand P, Blouin P. Improved detection of HBV DNA by PCR after microwave treatment of serum. Nucleic Acids Res 1991, 19: 4006.

Demeke T and Adams RP. The effect of plant polysaccharides and buffer additives on PCR. BioTechniques 1992, 12: 332-334

Grimprel E, Use of polymerase chain reaction and rabbit infectivity testing to detect Treponema pallidum in amniotic fluid, fetal and neonatal sera and cerebrospinal fluid. J Clin Microbiol 1991, 29: 1711-1718

Holodniy M, Kim S, Katzenstein D, Konrad M, Groves E, Merigan TC. Inhibition of human immunodeficiency virus gene amplification by heparin. J Clin Microbiol 1991, 29: 676-679

Khan G, Kangro HO, Coates PJ, Heath RB. Inhibitory effects of urine on the polymerase chain reaction for cytomegalovirus DNA. J Clin Path 1991, 44: 360-365

Klapper PE, Cleator GM, Tan SV, Guiloff RJ, Scaravilli F, Ciardi M, Aurelius E, Forsgren M. Diagnosis of herpes simplex encephalitis with PCR. The Lancet 1993, 341:691

Picard C, Ponsonnet C, Paget E, Nesme X, Simonet P. Detection and enumeration of bacteria in soil by direct DNA extraction and polymerase chain reaction. Appl Environ Microbiol 1992, 58:

2717-2722

Ruano G, Pagliaro EM, Schwartz TR, Lamy K, Messina D, Gaensslen RE, Lee HC. Heat-soaked PCR: an efficient method for DNA amplification with applications to forensic analysis. BioTechniques 1992, 13: 266-574

Sen S, Rahl S, Freireich EJ, Hewitt R, Stass SA. Detection of extrachromosomal circular DNA sequences from tumour cells by alkaline lysis, Alu-polymerase chain reaction technique. Mol Carcinog 1992, 5:107-110.

Singer-Sam J. The use of Chelex to improve the PCR signal from a small number of cells. Amplifications 1989, 3:11

Soini H, Skurnik M, Liippo K, Tala E, Viljanen MK. Detection and identification of mycobacteria by amplification of a segment of the gene coding for the 32-kilodalton protein. J Clin Microbiol 30: 2025-2028

Walsh PS, Metzger DA, Higuchi R. Chelex 100 as a medium for simple extraction of DNA for PCR-based typing from forensic material. BioTechniques 1991, 10: 506-513

DETECTION OF AMPLIFIED ONCOGENES BY DIFFERENTIAL POLYMERASE CHAIN REACTION

B. Brandt[1], U. Vogt[1/2], C. Griwatz[1/2], F. Harms[1/2], K.S. Zänker[2]

[1]Institute of Clinical Chemistry, WWU-Münster, Germany. [2]Institute of Immunology, University Witten-Herdecke, Witten, Germany

Genetic alterations such as oncogene amplifications have been found in a variety of human cancers and may have a prognostic impact. Therefore we used a non-radioactive tool for possible routine screening of genetic alterations with high fidelity in human cancer probes. This technique for detecting oncogene amplifications is a new application of the recently published method by Frye et al., called differential polymerase chain reaction (PCR). With this PCR we amplified five oncogenes which are involved in breast tumor metabolism (erb B-1, erb B-2, erb B-3, c-myc and Ha-ras). The level of amplification is reflected by the ratio of the oncogene and the single-copy reference genes (ß-globin, phenylalanine-hydroxylase) in gel electrophoresis. In a study of 169 breast cancer tissues, we showed that differential PCR is able to detect at least three fold oncogene amplifications. We regarded an oncogene as amplified if the cut off level of three fold amplification was passed. In about 40% of the carcinoma tissues we detected gene amplifications of at least one oncogene.

INTRODUCTION

Oncogenes were first identified as genes carried by RNA tumor viruses (Bishop et al., 1983). Yet it took a long time until homologous genes (proto-oncogenes) in non-transformed human tissue were detected. Proto-oncogenes are important for regulation of cell growth and differentiation (Alitalo et al., 1985). Genetic alterations such as gene amplifications, rearrangement of genes, point mutations and the loss of tumor suppressor genes (anti-oncogenes) can activate or change proto-oncogenes to oncogenes. In clinical studies oncogene amplifications, especially of the EGF-binding family, proved to be of prognostic importance in breast cancer. The transition of a proto-oncogene into an oncogene reveals biological properties of a cell that are involved in tumorigenesis (Whyte et al., 1988, Botchan et al., 1986).

Methods in DNA Amplification, Edited by
A. Rolfs *et al.*, Plenum Press, New York, 1994

FUNCTION OF DIFFERENT ONCOGENES

erb B-1 (EGF-R)

The receptor for EGF (epidermal-growth-factor) is a 170 kDa glycoprotein which is found on a variety of epithelial cells and is a part of the intracellular signal transduction pathway which regulates cell growth. Amplifications of EGF-R are accompanied by increasing levels of m-RNA and p170erbB-1 in cell lines and primary carcinomas. Therefore, elevated gene copy numbers of the EGF-receptor are probably important for tumor growth and prognosis (Wells, 1989).

erb B-2 (Her-2)

The erb B-2 oncogene was first identified in a chemically induced neuroblastoma in rats. The p185erbB-2 is a transmembrane receptor and one of the members of the family of the EGF binding proteins, but it is distinct from the p170erbB-1. Like EGF-R the erb B-2 protein has a carboxy-terminal moiety with tyrosine-kinase activity and an amino-terminal domain which is rich in cysteine residues. The human erb B-2 gene is an important prognostic factor in breast cancer, because oncogene amplifications were found in 10-40% of all breast carcinomas. Furthermore, p185 overexpression is correlated with poor prognosis. The correlation between clinical outcome and p185erbB-2 levels is much stronger in patients with lymph node metastasis compared to node-negative patients (Bargman et al., 1986, Slamon et al., 1987).

erb B-3 (Her-3)

Erb B-3 is another member of the EGF-R family. It was first isolated as a cDNA clone from a human carcinoma cell line by low-stringency hybridization. There is a large degree of conserved regions between the cytoplasmic domain of erb B-1 and erb B-2 but erb B-3 is less homologous to erb B-1 and erb B-2. The 6 kb mRNA of erb B-3 leads to a 160 kDa cell surface glycoprotein which is a part of the signal transduction pathways (Kraus et al., 1989, Plowman et al., 1990).

c-myc

The myc oncogene was first identified as a viral sequence of the MC29 avian retrovirus and is evolutionarily conserved. In human Burkitt¦s lymphoma and mouse plasmocytoma the coding sequences from the normal c-myc gene are translocated from chromosome 8q24 into a transcriptionally active immun globulin locus on chromosome 14. Gene amplifications of the c-myc oncogene have been found in several types of human carcinomas of the lung, colon and breast (Lee, 1989).

Ha-ras

The human ras gene family consists of three members (H-, K- and N-ras) encoding proteins with 85% homology. These p21 kDa glycoproteins are membrane associated, important for signal transduction, and a part of the superfamily of GTP-binding proteins. The human Ha-ras was identified as a homologue of the viral Ha-ras sequence isolated from Harvey rat sarcoma viruses. Human proto-oncogenes of the ras family are activated by point mutations within codon 12, 13 and 61 and other genetic alterations such as gene amplifications (Borresen, 1992). Such transformed ras oncogenes have been isolated from cell lines and human tumors and demonstrate the relevance of Ha-ras to human disease and prognosis (Der, 1989).

MATERIAL AND METHODS

The human breast carcinoma cell line (MCF-7) was obtained from the American Type Culture Collection (Rockville, Maryland, USA). DNA was isolated from i) the MCF-7 cell line, ii) from formalin-fixed, paraffin-embedded tissue sections and iii) from frozen tumor tissue by using the IsoQuick Nucleic Acid Extraction Kit (MicroProbe Corporation, Garden Grove, CA, USA). The gel-electrophoresis and silver-salt staining of PCR products were performed with the Separation/ Control Unit and the Development Unit of the PhastSystem from Pharmacia (Freiburg, Germany). The ratio of the stained signal intensities of the target gene to the single copy reference was evaluated and digitally scanned. The primers were made by gene map studies and synthesized with an automated DNA synthesizer (Pharmacia).

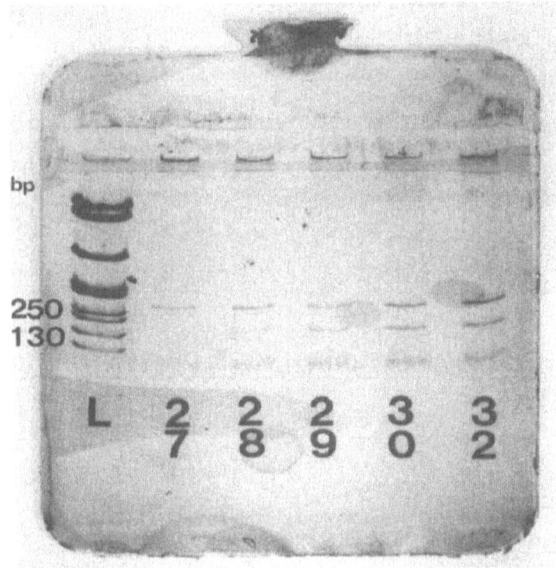

Figure 1. Quantitative evaluation of differential PCR with an optimal cycle number between 29-32 cycles.

PCR conditions

The mixture of differential PCR contains 100 ng to 1 µg of target DNA, 200 mM of each dNTP, 0.5 µM of each primer, 1U of Taq polymerase (Serva M1862, Heidelberg, Germany) and 1.5 µl of Taq polymerase buffer (50 mM Tris/HCl, ph 8.0/ 0.1 mM EDTA/ 100 mM KCl/ 1 mM DTT/ 50 % Glycerin/ 0,5 % Nonidet P-40 and 0.5 % Tween 20) in a total volume of 50µl with a 30µl mineral oil overlay. Differential PCR includes 29-32 cycles and 1 min at each

Table 1. Schematic flow-sheet of the experimental design

STANDARD PROTOCOL OF DIFFERENTIAL PCR

DNA-ISOLATION (30 min)

Samples: cell lines, tumor tissue and formalin fixed human carcinomas

DIFFERENTIAL PCR (2 h)

total volume: 50 µl
DNA: 100 ng-1 µg
Primer: 0,25 M each primer
dNTP′s: 200 mol each NTP
Taq Polymerase: 1U
Cycles: 29-32

DENATURATION: 1 min, 94°C
ANNEALING: 1 min, 62°C
EXTENSION: 1 min, 72°C

GEL ELECTROPHORESIS AND SILVER SALT STAINING (1,5 h)

20 % polyacrylamid gel

EVALUATION (30 min)

digital scanning

temperature (denaturation: 94°C/ annealing: 62°C/ extension: 72°C) except the initial cycle (4 min at 94°C). The final step was performed for 4 min at an extension temperature of 72°C on the RoboCycler 40 (Stratagene, Heidelberg, Germany).

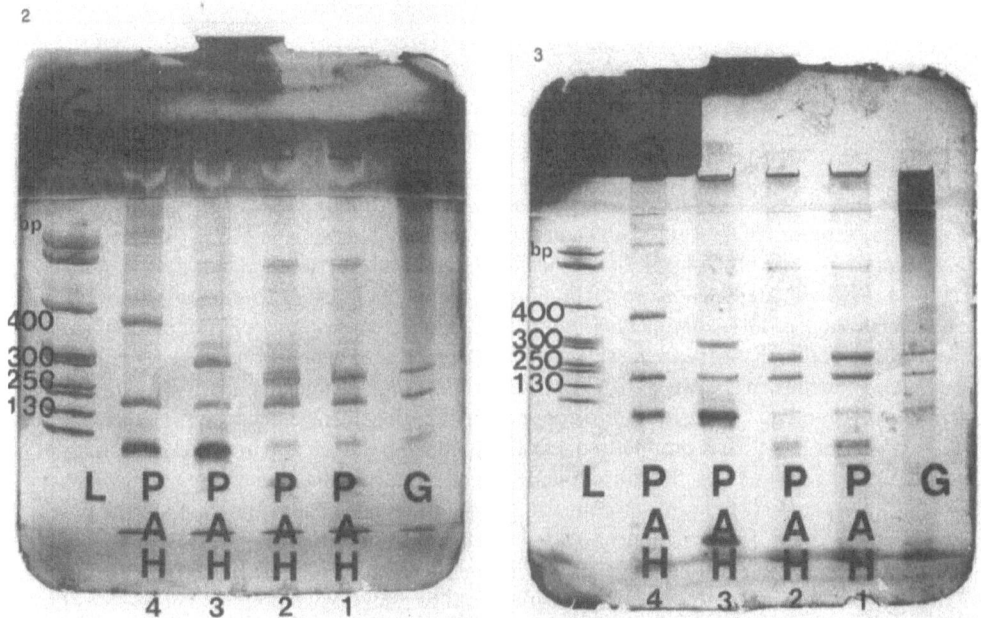

Figure 2-3. Two different single-copy reference genes (ß-globin, phenylalanine hydroxylase) to detect one-fold gene amplification of EGF-R in the MCF-7 cell line (G) and DNA isolated from healthy human donors (K).

We used one incubation cup containing the primer for the single copy ß-globin reference gene and obtained a 250 bp product; the primers for the test oncogenes were 25-30 bp in length and produced 130 bp products. The experimental conditions for both primer pairs were kept constant, e.g. cycles, annealing temperature, buffer, DNA and nucleotide concentration. 4 µl of each sample was electrophoresed on a 20% PAA-gel and the products were detected by silver salt staining. The densities of the PCR signals of the oncogene and the single-copy reference genes were digitally scanned at 600 dots per inch (dpi).

RESULTS

If there is no amplification the density values of the test gene products versus the single copy reference gene product are similar. On the other hand, if the ratio of densities is increased (above three), the oncogene is amplified. We used DNA of the MCF-7 cell line and showed that differential PCR is independent from the amount of template DNA in the range of 0.1µg to 1.0µg. At an annealing temperature of 62°C the ratio of amplification for erb B-1, erb B-2, erb B-3 and Ha-ras is almost 1 (no amplification). Interestingly c-myc is amplified two-fold under these experimental conditions (data not shown). After varying the annealing temperature in our experimental setting we could show that differential PCR works with high fidelity in the range of 50°C to 64°C. Beyond this window PCR cannot quantitatively detect genetic

Table 2. Annealing temperature and ratio of signal intensities in differential PCR

°C	48°C	50°C	55°C	60°C	62°C	64°C	66°C
Ratio:							not
EGF-R/	4.0	1.0	0.9	1.1	1.2	1.3	detected
ß-globin							

alterations, which might be due to nonspecific binding of the primers to the template. For example the ratio of EGF-R (erb B-1) and the single copy reference in MCF-7 breast cancer cells is only 1 (no amplification), if an annealing temperature between 50°C to 64°C is used (Table 2). We evaluated optimized cycle numbers of 29-32 in order to quantify the ratio of intensities of the two DNA fragments (Figure 1). Below 29 cycles we could not determine the accurate ratio of the oncogene to the single-copy reference gene because under these conditions the reaction kinetics of differential PCR is not in the range of linearity for both genes. For further verification we used four primers specific for the phenylalanine hydroxylase gene (PKU). The four PAH-PCR product sequences varied from 250 to 400 bp in length. PAA-gel electrophoresis revealed that the ratio of densities of the target oncogenes (at any given base pair length) were nearly the same as compared to the reference genes. The ratios of the EGF-R test gene and the PAH-references varied from 1.0 - 1.3, implying that EGF-R is not amplified in the MCF-7 cell line. We also controlled the non-amplification of erb B-1 in the MCF-7 cell line by using control DNA isolated from healthy human donors. The ratio of the EGF-R oncogene and the single copy PAH-reference gene was similar to the results we obtained from breast cancer cells. No amplification of EGF-R was detected when MCF-7 or control DNA was used, because the ratio of the oncogene PCR fragment and the single copy reference was approximately 1 (Table 3, Figure 2-3). In our study with 169 breast cancer tumor tissues we showed that oncogene amplifications are frequent events in tumorigenesis (Table 4). Each oncogene was found to be amplified in 10-20% of the samples and co-amplifications were seen as rare events (2.5%). Investigations of breast carcinomas revealed oncogene amplifications of at least one oncogene (not shown) in about 40% of the samples.

Table 3. Exchangeable single copy references without influencing the ratio of the reference gene and the test gene staining intensity by using DNA from the MCF-7 cell line and DNA isolated from healthy human donors.

MCF-7	**Ratio (Q)**	**bp of PCR products**
EGF-R/ß-globin	1.0	250
EGF-R/PAH1	1.0	250
EGF-R/PAH2	1.2	250
EGF-R/PAH3	1.2	300
EGF-R/PAH4	1.0	400
Healthy Donor	**Ratio (Q)**	**bp of PCR products**
EGF-R/ß-globin	1.2	250
EGF-R/PAH1	1.0	250
EGF-R/PAH2	1.1	250
EGF-R/PAH3	1.0	300
EGF-R/PAH4	1.3	400

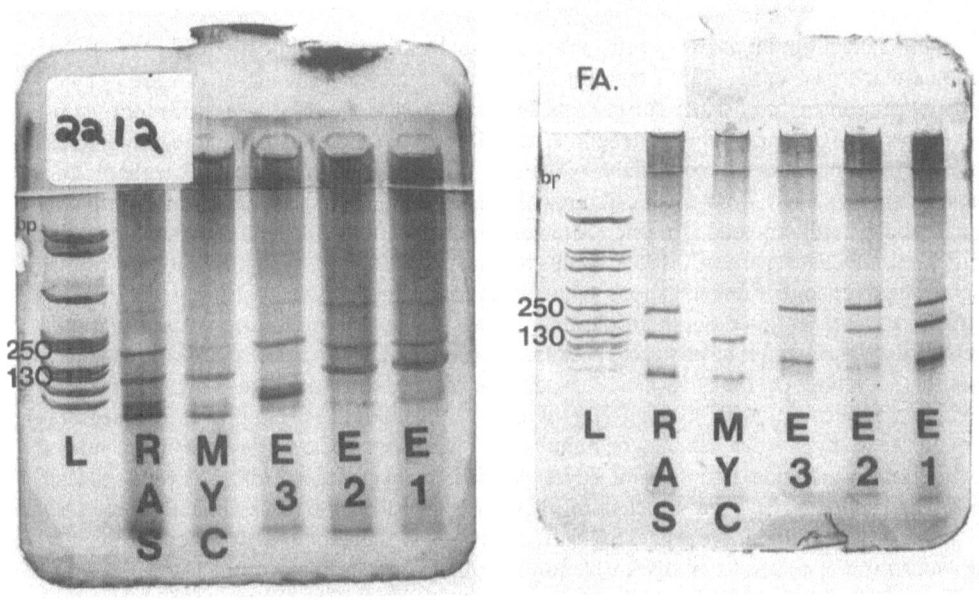

4

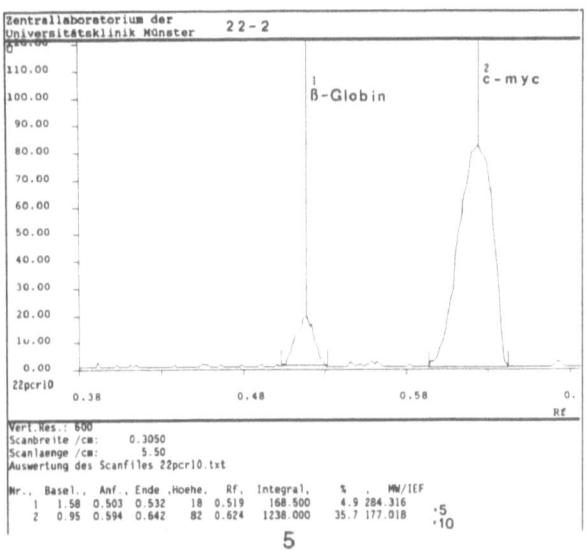

5

Figure 4 and 5: 5 to 10-fold amplification of the c-myc oncogene in human mammary carcinoma (left) and detection of a 30-fold c-myc oncogene amplification (right) in a fibroadenoma (FA). **Figure 6** (below) demonstrates the 5- to 10-fold amplification of c-myc from figure 4 in the digital scanning.

DISCUSSION

Many human tumors show genetic alterations such as gene amplifications, e.g. erb B-1 in glioma (Liberman et al., 1985) and erb B-2 in breast cancer (Slamon et al., 1987) which implies prognostic importance for patients survival. In PCR we have a sensitive tool for the detection oncogene amplifications (Saiki et al., 1985, 1986) and other genetic rearrangements like ras point mutations in epithelial malignancies (Lee et al., 1987). Also PCR technology is able to amplify DNA from small cell numbers (Saiki et al., 1988) as well as from formalin-fixed and paraffin-embedded tissue (Almoguera et al., 1988). We could demonstrate that differential PCR (Frye et al., 1989, Neubauer et al., 1992) is an adequate tool for routine screening of oncogene amplifications in human tumors with high fidelity. Differential PCR has several advantages over Southern blotting. When the Southern blotting is used, paraffin-embedded tissue cannot be analyzed, because degenerated DNA fragments are not reliable for Southern blotting. On the other hand reproducible products can be obtained by PCR within a template concentration of 100 ng-1µg. Even sensitive Southern blotting is not sensitive enough to analyze small samples of tumor tissue like fine needle biopsy. For conventional hybridization methods 2 to 10 g of DNA, which is equivalent to 2.000.000 cells, is used, whereas 100 tumor cells are sufficient for PCR analysis. Moreover, differential PCR does not require the use of radioactive probes for high sensitivity. Investigation of oncogene amplifications in human tumor tissue from DNA isolation to evaluation can be performed within 4.5 hrs, whereas with the Southern blotting the results are revealed after 4 days. Minimizing DNA quantity and time saving are not the only advantages of differential PCR. The costs of one Southern blot is about $ 130, in contrast to differential PCR which is 20-fold less expensive. In the future differential PCR may open new ways for tumor classification and patient prognosis because this technique is able to detect oncogene amplifications in benign and malignant tumor tissue. We showed that differential PCR is a useful technique to detect relative amplification grades under controlled and optimized conditions in normal and transformed DNA (Figure 4 and 6). Interestingly, the method may be able to detect amplifications in cells which might be on a transformation pathway (Figure 5: a benign fibroadenoma of the breast with a 30-fold c-myc amplification). If further results substantiate this finding, then differential PCR may be able to identify patients with benign adenofibromatosis who are at high risk of developing breast cancer (because oncogene amplifications are present). Thus, differential PCR may become a tool to screen for alterations which lead to a poorer prognosis in patients conventionally regarded as being at low risk.

Table 4. Prevalence of oncogene amplifications

oncogene	amplification	no amplification
erb B-1	11%	89%
erb B-2	14%	86%
erb B-3	19%	81%
c-myc	18%	82%
Ha-Ras	15%	85%

REFERENCES

Alitalo K, Bishop JM, Smith EJ, Chen EJ, Colby WW, Levinson AD. Nucleotide sequence of the v-myc oncogene of avian retrovirus MC29. Proc Natl Acad Sci USA 1983, 80: 100-104

Almoguera C, Shibata D, Forrester K, Martin J, Arnheim N, Perucho M. Most human carcinomas of the exocrine pancreas contain mutant c-K-ras genes. Cell 1988, 53: 549-554

Bargman CI, Hung MC, Weinberg RA. The neu oncogene encodes an epidermal growth factor receptor-related protein. Nature 1986, 319: 226-230

Bishop JM. Viral oncogenes. Cell 1985, 42: 23-38

Borresen A L. Oncogenesis in ovarian cancer. Acta Obstet Gynecol Scand 1992, Suppl, 25-30

Der CJ. Oncogenes. Kluwer Academic Publishers 1989, 73-120

Frye RA, Benz CC, Liu E. Detection of amplified oncogenes by differential polymerase chain reaction. Oncogene 1989, 4: 1153-1157

Kraus JH, Issing W, Miki T, Popescu NC, Aaronson SA. Isolation and characterization of erb B-3, a third member of erb B/epidermal growth factor receptor family: evidence for overexpression in a subset of human mammary tumors. Proc Natl Acad Sci USA 1989, 86: 7792-7796

Lee WMF. Oncogenes. Kluwer Academic Publishers 1989, 37-72

Lee MS, Chang KS, Cabanillas F, Freircich EJ, Trujillo JM. Science 1987, 237: 175-178

Liberman TA, Nusbaum HR, Razon N, Kris R, Lax J, Schlessinger J. Amplification, enhanced expression and possible rearrangement of EGF receptor gene in primary human brain tumours of glial origin. Nature 1985, 313: 144-147

Neubauer A, Neubauer B, He M, Effert P, Iglehardt D, Frye RA, Liu E. Analysis of gene amplification in archival tissue by differential polymerase chain reaction. Oncogene 1992, 5: 1019-1025

Plowman GD, Whitney G, Neubauer MG, Green JM, McDonald VL, Todaro GJ, Shoyab M. Molecular cloning and expression of an additional epidermal growth factor receptor related gene. Proc Natl Acad Sci USA 1990, 87: 4905-4909

Saiki RK, Scharf S, Faloona F, Mullis KB, Horn GT, Ehrlich HA, Arnheim N. Enzymatic amplification of ß-globin genomic sequences and restriction site analysis for diagnosis of sickle cell anemia. Science 1985, 230: 1350-1354

Saiki RK, Bugawan TL, Horn GT, Mullis KB, Ehrlich HA. Analysis of enzymatically amplified ß-globin and HLA-DQα DNA with allele-specific oligonucleotide probes. Nature 1986, 324: 163-165

Saiki RK, Gelfand DH, Stoffel S, Scharf SJ, Horn GT, Mullis KB, Ehrlich HA. Primer-directed enzymatic amplification of DNA with a thermostable DNA polymerase. Science 1988, 239: 487-491

Slamon DJ, Clark GM, Wong SG, Levin WJ, Ullrich A, McGuire WL. Human breast cancer; correlation of relapse and survival with amplification of the Her-2/neu oncogene. Science 1987. 235, 177-182

Wells A. The epidermal growth factor receptor and its ligands. Kluwer Academic Publishers 1989, 143-168

RESOLUTION OF A SECONDARY STRUCTURE IN AN UNKNOWN mRNA 5' END

M. Bacher, P. Hofmann and D. Gemsa

Institute of Immunology, Philipps University, Marburg, Germany

INTRODUCTION

Obtaining the complete cDNA is of critical importance for gene structure and expression studies. cDNA clones derived from libraries frequently lack the full-length 5'end of the sense strand. The RACE (rapid amplification of cDNA ends) PCR technique first described by Frohman et al (1988) has proven to be a versatile and fast method for obtaining the complete 5'end of cDNAs.

We previously showed (Gong et al 1991) that influenza A virus infection of the murine macrophage tumor cell line PU5-1.8 was associated with the induction of an additional, high molecular weight (hmw) tumor necrosis factor (TNF)-α mRNA, whereas a bacterial stimulus such as lipopolysaccharide (LPS) alone triggered only the regular TNF-α mRNA. In order to characterize the hmw TNF-α mRNA species in macrophages we first performed an extensive Northern analysis covering the different structural elements of the cytokine gene locus. Based on the hybridization analysis it was most likely that the elongation of the additional cytokine mRNA was located at the 5'end. To further characterize the elongated 5'-end of the additional cytokine mRNA, we chose the RACE procedure as schematically illustrated in Figure 1. An antisense primer (GSP 1) designed of about 200 basepairs upstream from the regular transcription starting site was used to prime the first strand synthesis by use of reverse transcriptase. As mentioned in the original RACE protocol, a homopolymeric tail (dC) was added to the purified 3' end of the first strand product by use of terminal deoxyribonucleotide transferase. The homopolymeric (dC)-anchor served then as a defined sequence to which a complementary primer, containing an oligo(dG) stretch, annealed. To accomplish the first round of amplification, another PCR primer was designed slightly upstream from the first cDNA synthesis primer (GSP 1), and thus served as a so called „nested primer" for additional specificity. In order to visualize the 5' RACE products on a gel, it was essential to use another nested primer amplification. For this purpose we designed an antisense primer (GSP 3) covering the regular transcription starting site in order to amplify exclusively cDNAs which were elongated at this position.

Methods in DNA Amplification, Edited by
A. Rolfs *et al.,* Plenum Press, New York, 1994

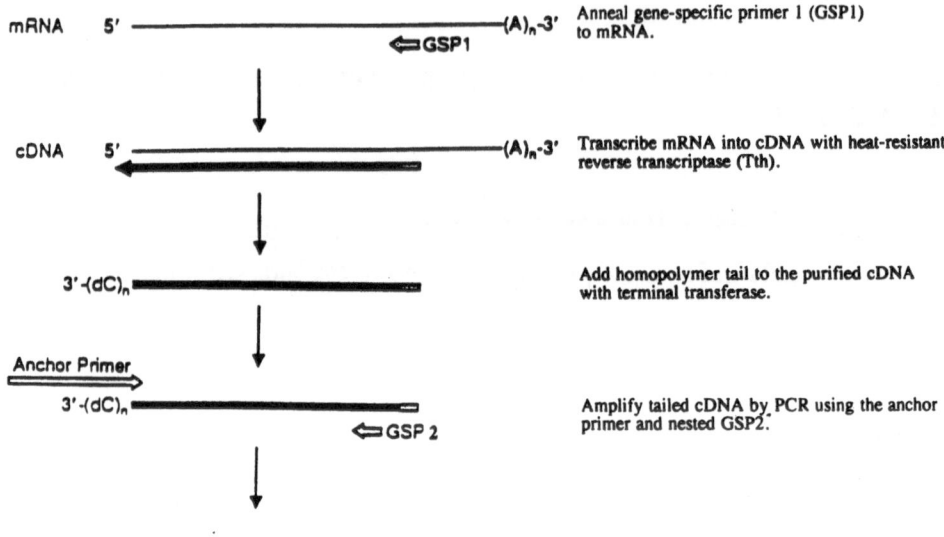

mRNA   5' ——————————————————(A)ₙ-3'   Anneal gene-specific primer 1 (GSP1)
                              ⇐GSP1   to mRNA.

cDNA   5' ——————————————————(A)ₙ-3'   Transcribe mRNA into cDNA with heat-resistant
                                                        reverse transcriptase (Tth).

3'-(dC)ₙ ■■■■■■■■■■■■■■■■■■■■■■■■■   Add homopolymer tail to the purified cDNA
                                                        with terminal transferase.

Anchor Primer
3'-(dC)ₙ ■■■■■■■■■■■■■■■■■■■■■■■■■   Amplify tailed cDNA by PCR using the anchor
                              ⇐GSP 2   primer and nested GSP2.

Second round of amplification with nested primer.

Figure 1. Schematic diagram of the 5'RACE method

MATERIALS AND METHODS

Macrophages

The murine macrophage cell line PU5-1.8 was propagated in vitro in RPMI 1640 medium supplemented with 5% heat-inactivated (56°C, 30 min) fetal calf serum (FCS), penicillin (100 U/ml), streptomycin (100 µg/ml), L-glutamine (2 mM), nonessential amino acids (2 mM), pyruvate (2 mM), and HEPES buffer (10 mM).

Virus preparation

Influenza A virus strain Puerto Rico 8 (H1N1), abbreviated A/PR8, was propagated in the allantoic cavity of 11-day old embryonated hen eggs for 48 hrs at 37°C and then kept for 12 hrs at 4°C. Harvest and purification of A/PR8 has been described in detail (Nain et al 1990; Klenk and Rott 1972).

Infection of Macrophages

After adherence to plastic dishes, 0.5×10^6 macrophages/ml were washed twice and inoculated with influenza A viruses at a multiplicity of infection (MOI) of 2 for 1 h under serum-free conditions. The infection period and subsequent incubation was performed in the presence or absence of LPS (E.Coli 0127:B8, Difco)

RNA preparation

8 hrs after exposure of PU5-1.8 macrophages to A/PR8 (2 MOI), in the presence or absence of LPS (2ng/ml), total RNA was isolated by a single step method previously described (Chomczynsky and Sacchi, 1987).

cDNA Synthesis

First strand cDNA was prepared according to instructions supplied with the Superscript II reverse transcriptase (Gibco BRL) or Tth DNA polymerase (Boehringer Mannheim).

Purification and dC-Tailing of cDNA

Purification and dC-tailing of the cDNA was performed according to instructions supplied with the 5'RACE kit (Gibco BRL).

PCR

For the PCR amplification of the 5'RACE products we incubated the DNA samples at 94°C for 20 min in 46.5 μl 1.1 x PCR buffer. A 3.5 μl bolus of concentrated deoxynucleotides, Taq

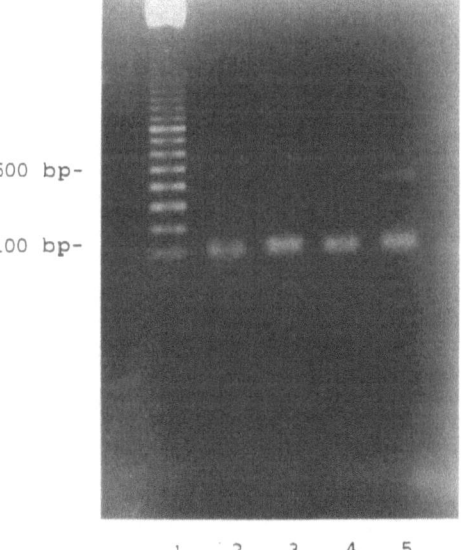

Figure 2. PCR nested primer amplification of the 5'end of the hmw TNF-α mRNA using Tth for the first strand cDNA synthesis. Total RNA was isolated from murine macrophage cell line PU5-1.8. 1 μg RNA was used as the template for the first strand cDNA synthesis The PCR reaction was performed in a volume of 50 μl, in the presence of 50 pM of each primer, 200 μM of each dNTP (Promega) in a buffer containing 10 mM Tris HCl pH 8.3, 50 mM KCl, 1.5 mM MgCl₂, 0.01% gelatin, and 4 units of Taq polymerase (Perkin-Elmer Cetus). Thermal cycling involved 30 cycles of: 1 min at 94°C; 1 min at 55°C; and 1 min at 72°C. Finally, 10 μl of each reaction was separated on a 2% agarose gel and stained with ethidium bromide. Lane 1: 100 bp DNA ladder (Pharmacia). Lane 2: Negative control without template. Lane 3: cDNA from untreated macrophages. Lane 4: cDNA from LPS-treated macrophages. Lane 5: cDNA from virus-infected macrophages (hmw TNF-α induction).

Polymerase (Perkin-Elmer Cetus) and primers prepared without buffer was then added prior to thermal cycling. This heat-soaked PCR has previously been described in detail (Ruano et al 1992).

PCR cycling conditions were as follows: 94°C for 1 min, 55°C for 45 s, 72°C for 1 min (30 cycles), followed by a 5 min final extension at 72°C. Nested amplification were initiated with 1 μl amplification products from the first round.

RESULTS

When the standard RACE strategy was performed, we only obtained truncated and weak bands instead of an expected amplification product of about 500 bp in the virus-infected sample.

Changing the cycle profile by increasing the annealing temperature (from 50 to 68°C), modifying the concentration of Mg^{2+} (from 1.5 to 4 mM) in the PCR buffer, or adding cosolvents like DMSO or glycerol (3 to 10%), previously demonstrated to improve PCR amplification of DNA with complex secondary structure (Smith et al 1990; Shen and Hohn, 1992), did not result in amplification of a fragment of the correct size.

However when we introduced a heat-resistant reverse transcriptase (Tth) in the basic RACE protocol, thus increasing the temperature from 42 to 70°C during the first strand synthesis, we eventually obtained a full length additional sequence elongation at the 5'end of the hmw TNF-α cDNA, which was in accord with the virus-induced hmw TNF-α mRNA.

Thus, these modifications of the original RACE procedure are most suitable for the resolution of excessive secondary structures frequently observed in the mRNA 5'end.

REFERENCES

Chomczynsky P and Sacchi N. Single step method of RNA isolation by acid guanidinium thiocyanate-phenol-chloroform extraction. Anal Biochem 1987, 162:156-159

Gong JH, Sprenger H, Hinder F, Bender A, Schmidt A, Horch S, Nain M, and Gemsa D. Influenza a virus infection of macrophages: enhanced tumor necrosis factor-α (TNF-α) gene expression and lipopolysaccharide-triggered TNF-α release. J Immunol 1991, 147:3507-3513

Klenk HD and Rott R. On the structure of the influenza virus envelope. Virology 1972, 47:579-588

Nain M, Hinder F, Gong JH, Schmidt A, Bender A, Sprenger H, and Gemsa D. Tumor necrosis-α production of influenza A virus-infected macrophages and potentiating effect of lipopolysaccharides. J Immunol 1990, 145:1921-1929

Ruano G, Pagliaro EM, Schwartz TR, Lamy K, Messina D, Gaensslen RE and Lee HC. Heat-soaked PCR: an efficient method for DNA amplification with application to forensic analysis. Biotechniques 1992, 13:266-274

Shen WH and Hohn B. DMSO improves PCR amplification of DNA with complex secondary structure. TIG 1992, 8:227

Smith KT, Long CM, Bowman B, and Manos M. Using cosolvents to enhance PCR amplification. Amplifications: A Forum for PCR Users 1990, 5:16-17

AUTOMATED SOLID-PHASE SEQUENCING OF GENOMIC DNA OBTAINED FROM POLYMERASE CHAIN REACTION

A. Rolfs and I. Weber

Working Group for Molecular Neurobiology, Institute for Neuropsycho-pharmacology, Free University, Berlin, Germany

INTRODUCTION

In the last few years DNA sequencing has become increasingly important to molecular biology and biotechnology for DNA analysis. Currently there is a attempt to determine the complete genome sequences of various species (e.g. the human genome project). Such research requires technically easy, automated procedures which produce reliable and reproducible results. For these large-scale sequencing projects to be succesful different systems and procedures have to be combined and integrated (Harrison et al 1993, Wilson et al 1988a, Wilson et al 1988b).

To perform DNA sequencing of genomic DNA obtained from PCR under optimal conditions it is necessary to obtain a well purified and single-stranded (ss) DNA template. Unincorporated primers and dNTP's have to be removed from the amplified DNA because they will impair the sequencing reaction. Furthermore, a ssDNA template avoids interference with the sequencing reaction due to reassociation of the double-stranded PCR product and leads to increased length and accuracy of readable sequences (Rolfs et al 1992); additionally, less DNA template is required than for dsDNA. A number of methods for generating ssDNA have been described: 1. Generation of ss DNA by asymmetric PCR (Gyllensten and Erlich 1988, Wilson et al 1990); 2. generation of ssDNA by blocking-primer PCR (Gyllensten 1989); 3. exonuclease digestion of the phosphorylated strand of the ds DNA (Higuchi and Ochman 1989); 4. immobilization of amplified biotinylated dsDNA by avidin and strand-specific elution using alkali: a) affinity agarose gel (Mitchell and Merril 1989, Stahl et al 1988), b) capturing biotinylated PCR product using streptavidin-coated paramagnetic beads (Syvänen et al 1991).

The disadvantage of ssDNA sequencing is that the generation of ssDNA is associated with additional steps prior to the sequencing reaction and is thus more time-consuming. Also, most procedures lead to a loss in the yield of the DNA template. Furthermore, the isolation of

Methods in DNA Amplification, Edited by
A. Rolfs *et al.*, Plenum Press, New York, 1994

ssDNA from a strand-separating gel involves a higher risk of cross-contamination with other PCR products.

We present a protocol for DNA-sequencing using a solid-phase method with streptavidin-coated paramagnetic beads; the protocol has the advantage that it can be performed as a fully automated process followed by the sequencing reaction. Several methods for fully automated DNA sequencing procedures have already been described (Koop et al 1990, Koop et al 1993, Wilson et al 1990, Wilson et al 1988). The problem with these methods is the fact that they disregarded a cost-effective and fast strand separation method. The use of a 96-well microtiter plate is recommended for automated laboratory work because most robotic workstations in the routine lab are adapted to that format.

Any device used for the automated separation of DNA containing one biotinylated strand has to meet the following requirements: 1. the production of a strong and homogenous magnetic field in all 96 wells of the microtiter plate, 2. agitation, in order to resuspend the paramagnetic beads following separation steps, 3. temperature control up to 96°C and 4. easy adaptability to an existing pipetting robot. Furthermore, the microtiter plate must be able to be placed directly onto the heating/magnetic block to ensure a satisfactory heat transfer between the heating block and the sample.

Here we describe an automated solid-phase protocol for direct sequencing of genomic PCR products, using magnetic beads coated with streptavidin as solid support and an instrument (PolySeq™) with heating, magnetic and mixing functions which is integrated in a robotic workstation (Biomek™ 1000, Beckman Instruments, Fullerton, CA). This solid-phase method is extremely useful for template purification and strand separation of DNA obtained from polymerase chain reaction (PCR). Compared with dsDNA sequencing, this method yields consistently better results with increased lengths of readable sequences and reduced backgrounds. For the sequencing reaction we used a non-radioactive dideoxynucleotide chain termination procedure (Sanger et al 1977) with fluorescent-dye-labelled primers (Knight 1988). A similar protocol for the semi-automated application of bidirectional solid-phase sequencing of plasmid DNA was published by Hultman and coworkers (Hultman et al 1991). This group also used a Biomek™ 1000 robotic workstation equipped with a heating/cooling unit but without an instrument for the automated separation of the paragmagnetic beads.

METHOD AND MATERIALS

Primer design

The biotinylated PCR product is produced using a 5'-biotin-labeled amplification primer. The primers can be synthesized on a Millipore DNA-synthesizer (Cyclone™); the biotinylation is done by coupling biotin-phosphoramidite at the 5'- end during the last synthesis reaction step. The forward primer (P1) contains the specific amplification sequence [for the 5SrRNA Legionellae gene 5' ACT-ATA-GCG-ATT-TGG-AAC-CA 3'] and a 5'-sequence identical to that of the M13 universal fluorescent sequencing primer [5'TAA-AAC-GAC-GGC-CAG-TGC-CA 3']. The non-biotinylated reverse primer (P2) contains the specific sequence of the 3'-primer [for the 5SrRNA Legionellae gene 5' GCG-ATG-ACC-TAC-TTT-CGC-AT 3'] and a 5'-sequence identical to that of the universal reverse M13 fluorescent sequencing primer

[5'CAG-GAA-ACA-GCT-ATG-ACC 3']. For PCR the primers are used in equimolar ratio. After PCR, 40μl of the specific biotinylated amplification product is pipetted into a 96-well microtiter plate. Figure 1 demonstrates the procedure for producing a biotinylated PCR amplicon.

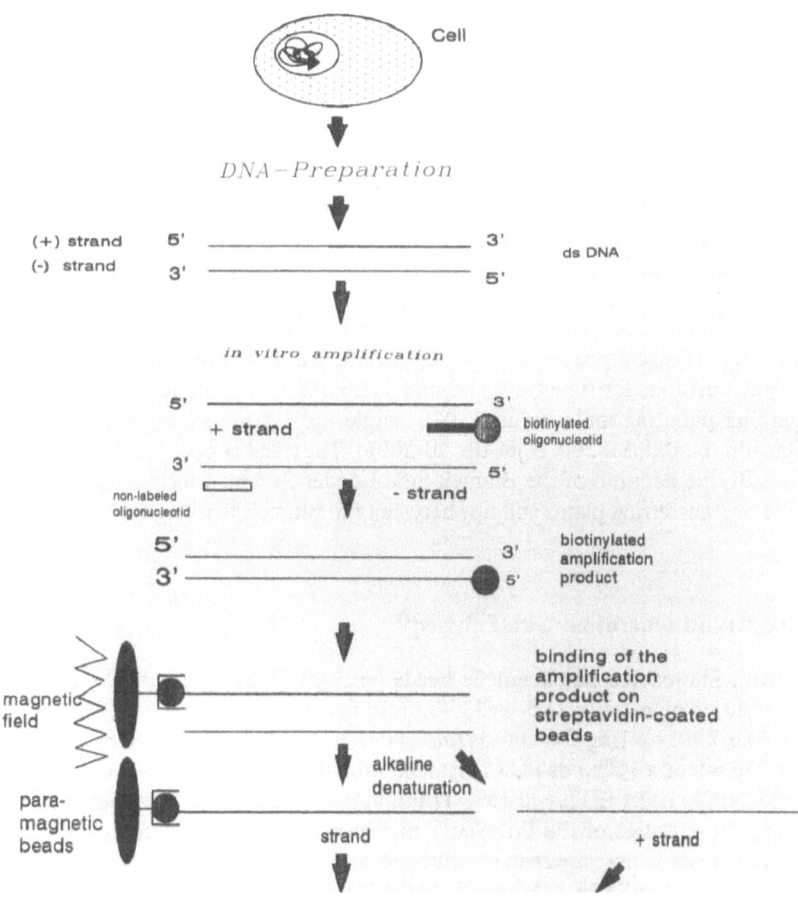

Figure 1. Schematic flow-sheet demonstrating the procedure for producing a biotinylated PCR product followed by the strand separation with streptavidin coated paramagnetic beads using the PolySeq™ and Biomek™ 1000.

Construction of the automated magnetic station (PolySeq™)

PolySeq™ was designed by us and constructed by Polygen Corp. (Frankfurt, FRG); the robotic workstation (Biomek™ 1000) which carries out the pipetting steps is from Beckman Instruments (Fullerton, CA). The current version of PolySeq™ consists of a block with integrated temperature, magnet and agitation unit as well as a BIOMEK software-linked controller. The heating block can accomodate a 96-well microtiter plate with a U-bottom (flexible plates from Falcon™, Becton Dickinson) and has a regulation precision of ±0.2°C of temperatures between 30°C to 98°C. The heating block has an overall temperature uniformity of ±0.4°C. The heating performance is 1°C/sec. The cooling is done by an built-in ventilator and the cooling performance is about 0.8°C/sec. The heating block can be agitated with a stroke of ±1mm. Length and frequency of agitation can be regulated by the software. Agitation causes the uniform redistribution of paramagnetic beads after separation has taken place. The maximum sample volume which can be agitated without spilling is about 200µl. PolySeq™ also has a platform with several neodymium-iron-boron magnets (Neo-Delta™, IBS Magnet, FRG) in the form of a narrow strip; these magnets can fit between the wells of the microtiter plate (Figure 4) and attract the paramagnetic beads to the sidewalls of the wells. The platform is elevated to separate the beads (which occurs within 5-15 seconds) and is then lowered so that the beads can be resuspended in the next reaction step. The problem of evaporation occuring during the reaction in the uncovered microtiter plates and particularly during the heating steps is solved by adding deionized, distilled water (usually 2µl) after the annealing step. Moreover, during the heating steps we cover the microtiter plate with a flexible lid which can be moved by the Biomek Side Loader.

The Biomek™ 1000 automated workstation (Frank et al 1988) consists of a holder for a tip rack, two tray positions for tube trays and one position for the PolySeq™. For the pipetting procedures four pipetting tools are used: P20, single-tip, 1-20µl; P200, single-tip, 20-200µl; MP20, eight-tip, 1-20µl; MP200, eight-tip, 20-200µl. The transfer operation can be completely automated with the addition of the Biomek Side Loader System which consists of a robotic arm capable of transferring plates and tips between the Biomek working table and the storage area.

Automated strand separation with PolySeq™

40µl streptavidin-coated paramagnetic beads per well (M280 Streptavidin Dynabeads™, Dynal AS, Oslo) is pipetted in a 96-well U-bottom flexible microtiter plate (Falcon™ Becton Dickinson, Lab. #3911). 1mg Dynabeads (ca. 100µl, 6-7 x 10^8 beads per ml dissolved in PBS) bind about 25pmol of a 300bp ds PCR fragment. 40µl of binding buffer (10mM Tris-HCl, pH 7.5, 100mM NaCl, 1mM EDTA, 0.15% Triton X100) is added for the equilibration of the beads. Using the magnets of the PolySeq™ platform the beads were completely separated (usually about 40sec), the supernatant removed and discarded. After the beads had been washed twice, 27µl 20xSSPE (3M NaCl, 2mM NaH_2PO_4 x H_2O, 2mM EDTA) plus 40µl biotinylated PCR product are added and then well shaken by the agitation function of PolySeq™. The reaction was incubated for 30min at 37°C with gentle agitation steps every two minutes. During the incubation, the biotinylated PCR products are bound to the beads with streptavidin bridges. The beads can be separated again by the magnets. After removing and discarding the supernatant 50µl of denaturation buffer (0.1M NaOH, 0.04mM EDTA) is added and the beads resuspended by shaking and incubated for 3min at room temperature. After this incubation the beads are separated again, the supernatant removed, stored in a new reaction well and neutralized with 5µl 1M HCl. The supernatant contains the non-biotinylated strand and is stored for further sequencing. After the magnets have been removed the beads are washed with 200µl binding buffer and finally resuspended in 16µl water and are now ready for the sequencing procedure.

Sequencing reaction

Except for the M13 reverse primer the sequencing reagents for the following solid-phase sequencing are from the Sequenase™ Dye-Primer Sequencing-Kit (Applied Biosystems, #401117). The kit contains 80pmol of the "A"- and "C"-specific primers JOE and FAM, plus 160pmol of the "G"- and "T"-specific primers, TAMRA and ROX. The reaction procedures for the sequencing assay correspond to the manufacturers recommendations. All pipetting steps required for the protocol are programmed with the Biomek™ 1000 software. Twenty-four sets of reactions were carried out in 96-well microplates.

The non-biotinylated minus-strand is precipitated by 95% ethanol, the pellet resuspended in 6µl deionized distilled water and used for the Taq sequencing procedure (Taq Dye Primer Cycle Sequencing Kit, -21M13, Applied Biosystems, #401119). The pipetting steps for the sequencing reaction are also done by the robotic workstation. We do not presently employ cycle sequencing when sequencing the nonbiotinylated strand with Taq. The chain-termination reactions are performed at a temperature of 70°C for 15min. The annealing step is similar to the procedure with Sequenase™: 5 min at 65°C followed by slow cooling to room temperature within 10 min. Although the pipetting steps are not performed at 4°C and cooling after the execution of the annealing and termination steps is done only to room temperature, no significant difference in sequencing quality has been observed.

RESULTS

Quantification of PCR products necessary for automated pipetting and sequencing procedures

We have amplified a 144 base pair (bp) DNA fragment (Figure 2) from the 5SrRNA gene from Legionella pneumophila (ATCC 43110 Oxford) as well as a 501bp fragment from the envelope region of HIV-1 using specific oligonucleotides modified with the M13 and biotinylated reverse-M13 sequence as described in methods. For the automated sequencing procedure

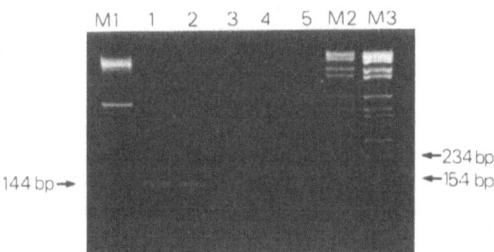

Figure 2. Agarose gel analysis of a 144bp PCR fragment (5SrRNA) from Legionella pneumophila (ATCC 43110 Oxford). 10µl, 7.5µl, 5µl, 2.5µl and 1.25µl (lane 1 to 5) of a 100µl PCR sample were separated in an ethidium bromide-stained 3% agarose/NusieveTM gel in 1xTBE-buffer. M1: 1µl 100bp-ladder (Bibco BRL), M2 and M3: 1µl and 2µl of the Boehringer VI marker respectively. Semiquantification of the PCR fragments reveals 60ng product in lane 1 and about 8ng in lane 5. 50-60ng is sufficient for sequencing using the robotic workstation.

with the PolySeq™ different amounts of the PCR fragments (1,000ng, 500ng, 250ng, 60ng, 30ng) were used for testing the smallest amount of the amplicon that can be sequenced according our procedure. As shown in figure 3 even amounts as little as 300ng and 60ng used in the automated sequencing procedures gave excellent quality in the sequencing data. No differences in sequencing quality were found using PCR fragment amounts up to 60ng. Below the cut-off point of about 60ng we were not able to obtain sufficient sequencing signals. Since the yield of an optimized 100μl PCR assay (35 cycles) ranges mostly between about 500ng and 1,500ng DNA it will be possible to do at least about five to ten sequencing reactions per PCR assay with our robot system.

Using the PolySeq™ for increasing sequencing capacity

The automation process described in the paper includes automated strand separation of dsDNA followed by automated sequencing of the separated single strands. With this method one can determine the DNA sequence of 24 350bp-PCR amplicons in 24 hours using a totally computer-controlled system. Using this system one person can comfortably analyze 30,000 to 40,000bp sequence data per week using the ABI 373A sequencing machine, with time left over for data analysis and other tasks.

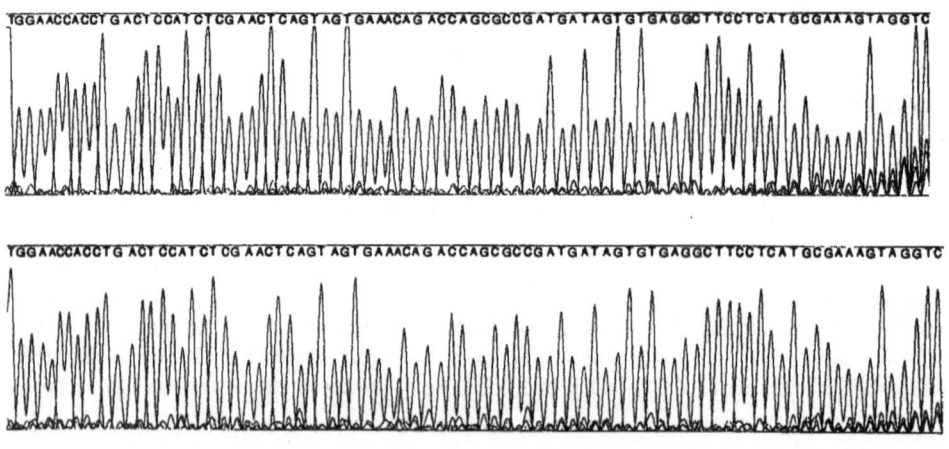

Figure 3. Demonstration of the Sequenase™ sequencing results of the single-stranded 144bp 5SrRNA fragment of Legionella pneumophila using different amounts of DNA template for sequencing reaction (Björn Heidrich). In the upper chromatogramm about 300ng template has been used, in the middle 60ng and in the lower 30ng, respectively. There are no obvious differences in quality between the upper and the middle chromatogram. Below the cut-off of about 50-60ng we were not able to get sufficient sequencing signals with the ABI 373A sequencing machine.

Sequence length and error analysis

Sequences for two PCR prodcuts (144bp 5SrRNA Legionella pneumophila and 501bp fragment from the env-HIV-1 gene) were determined as described above and were compared with manually determined reference sequences using the GeneWorks™ computer program. Thirtysix 144bp fragments and ninetyfour 501bp fragments were sequenced with Sequenase™ as well as with Taq polymerase on the robotic workstation. Analysis of these fragments showed an error rate of about 1% for Taq polymerase and about 0.3% for Sequenase™ (mismatch, insertion, deletion, ambiguity). After approximately the 300th base in a sequence the error frequency appears to increase; therefore the quality of the 144bp sequence was higher than that of the 501bp sequence with both manual and automated sequencing. With nucleotides which exceed a length of 350bp the number of sequences avaiable for comparison becomes too small and errors occur at a frequency of 6-20%. The differences in the level of errors obtained from regions beyond 350bp seem to depend mainly on the quality of the PCR assay.

Figure 4. Prototype of the PolySeq™ with heating block, agitation unit and the magnets at the top of the pins (arrow) - 1 per 4 wells - integrated in the Biomek robotic workstation.

DISCUSSION

The PolySeq™ is a new accessory for the Biomek™ 1000 that permits automated strand separation of dsPCR-generated DNA fragments as well as the sequencing reaction for both T7-polymerase and Taq polymerase. In contrast to prior systems (Hultman et al 1991) our method allows the complete automatisation process for single-stranded sequencing DNA obtained from polymerase chain reaction. The system configuration enables the automated sequencing reaction of about 30.000 - 40.000 base pairs per week using an ABI 373A sequencing machine and reduces the time necessary for the sequencing procedure by 50-70%.

Along with this higher efficiency time is gained in which a technician can analyze data and pipet the next PCR assay whilst reactions are underway. DNA strand separation and sequencing done with the robotic workstation achieves an accuracy of about 99.7% using the T7-polymerase and 99.0% using the Taq polymerase. These data agree with those from other groups doing their separation and sequencing procedures manually (Koop et al 1993). Due to the uniform peak height Sequenase™ seems to be ideal for detecting heterozygotic changes in the sequence. Nevertheless, we found no statistically significant differences in the ambiguity, rate of deletions, insertions and mismatches between the two enzymes used in the study.

The PolySeq™ achieves a highly uniform temperature distribution and has a regulation precision of ±0.2°C. Therefore, it is possible to design an automated assay including PCR, strand separation and sequencing. This means that an even greater increase in efficiency can be realized.

In summary, the fully automated sequencing method described in this paper affords highly standardized sample handling, eliminates pipetting errors, and allows secure and efficient sample processing.

ACKNOWLEDGEMENT

The authors would like to acknowledge the excellent technical assistance and advice from Mr. Roland Friedberger. We thank Dr. Peter Robinson for his critical reading of the manuscript.

REFERENCES

Frank R, Bosserhoff A, Boulin C, Epstein A, Gausepohl H, Ashman K. 1988. Automation of DNA sequencing reactions and related techniques: a workstation for micromanipulation of liquids. BioTechnology 1988, 6: 1211-1213

Gyllensten U B, Erlich H A. Generation of single-stranded DNA by the polymerase chain reaction and its application to direct sequencing of the HLA-DQA locus. Proc Natl Acad Sci USA 1988, 85: 7652-7656

Gyllensten UB. PCR and DNA sequencing. BioTechniques 1989, 7: 700-708

Harrison D, Baldwin C, Prockop D J. Use of an automated workstation to facilitate PCR amplification, loading agarose gels and sequencing of DNA templates. BioTechniques 1993, 14: 88-97

Higuchi RG, Ochman H. Production of single-stranded DNA templates by exonuclease digestion following the polymerase chain reaction. Nucleic Acids Res 1989, 17: 5865

Hultman T, Bergh St, Moks T, Uhlen M. Bidirectional solid-phase sequencing of in vitro- amplified plasmid DNA. BioTechniques 1991, 10: 84-93

Knight P. Automated DNA sequencers. BioTechnology 1988, 6: 1095-1096

Koop BF, Wilson RK, Chen C, Halloran N, SciammisR, Lindelin J, Hood L. Sequencing reactions in microtiter plates. BioTechniques 1990, 9:10-14

Koop BF, Rowan L, Chen WQ, Deshpande P, Lee H, Hood L. Sequenase length and error analysis of SequenaseR and automated Taq cycle sequencing methods. BioTechniques 1993, 14: 442-447

Mitchell LG, Merril CR. Affinity generation of single-stranded DNA for dideoxy sequencing following the polymerase chain reaction. Anal Biochem 1989, 178: 239-242

Rolfs A, Schuller I, Finckh U, Weber-Rolfs I. PCR: Clinical diagnostics and research. Springer Verlag, New York, Heidelberg 1992, pp23-67

Sanger F, Nicklen S, Coulson AR. DNA sequencing with chain-termination inhibitors. Proc Natl Acad Sci USA 1977, 74: 5463-5467

Stahl S, Hultman T, Olsson A, Moks T, Uhlen M. Solid phase DNA sequencing using the biotin-avidin system. Nucleic Acids Res 1988, 16: 3025-3038

Syvänen AC, Hultman T, Aalto-Setälä K, Söderlund H, Uhlen M. Genetic analysis of the polymorphism of the human apolipoprotein E using automated solid-phase sequencing. GATA 1991, 8: 117-123

Wilson RK, Chen C, Hood L. Optimization of asymmetric polymerase chain reaction for rapid fluorescent DNA sequencing. BioTechniques 1990, 8: 184-189

Wilson RK, Clark S, Yuen AS, Hood L. Automation of dideoxynucleotide DNA sequencing reactions using a robotic workstation. BioTechniques 1988a, 6: 776-787

Wilson RK, Clark S, Yuen AS, Hood L. Automation of dideoxynucleotide DNA sequencing reaction using a BIOMEK 1000 robotic workstation. FASEB J 1988b, 2: A1125

Alternative DNA-Amplification Methods

LIGASE-MEDIATED DETECTION TECHNIQUES

Martin Wiedmann[1], Wendy Wilson[2], John Czajka[1], Francis Barany[3] and Carl A. Batt[1*]

[1]Department of Food Science, Cornell University, Ithaca, NY, USA, [2]Department of Plant Pathology, Cornell University, Geneva, NY, USA, [3]Department of Microbiology, Cornell University Medical College, New York, NY, USA

INTRODUCTION

Since its introduction in 1985, the polymerase chain reaction (PCR) has revolutionized molecular biology and has facilitated the development of a variety of nucleic acid-based detection systems for genetic disorders as well as for bacterial, viral and other pathogens. Ligase chain reaction (LCR) is a recently developed DNA amplification technique, which utilizes a thermostable ligase and allows the discrimination of DNA sequences differing in only a single basepair. A measure of the intellectual genesis of LCR can be traced back to the oligonucleotide ligation assay (OLA) (Landegren et al 1988; Nickerson et al 1990). This method was used in conjunction with a primary PCR step to screen for sickle cell anemia and other human hereditary disorders. Compared to the OLA, the LCR provides a much higher degree of sensitivity and is less susceptible to false-positive ligation product formation. The use of a thermostable ligase minimizes target independent ligation since the ligation reaction can be performed at or near the melting temperature (T_M) of the oligonucleotide primers (Barany 1991a).

PRINCIPLE OF LCR

The principle of LCR is based upon the ligation of two adjacent oligonucleotide primers, which uniquely hybridize to one strand of the target DNA. The junction of the two primers is positioned such that the nucleotide at the 3' end of the upstream primer coincides with a potential single basepair difference in the targeted sequence, which defines the different alleles or species. If the target nucleotide at that site complements the nucleotide at the 3' end

Methods in DNA Amplification, Edited by
A. Rolfs *et al.*, Plenum Press, New York, 1994

of the upstream primer, the two adjoining primers can be covalently joined by a DNA ligase. In the LCR, a second pair of primers, complementary to the first pair is present, again with the nucleotide at the 3' end of the upstream primer denoting the sequence difference (Figure 1). In a cycling reaction, using a thermostable *Thermus aquaticus* DNA ligase, the ligated product can then serve as a template for the next reaction cycle leading to an exponential amplification process, analogous to PCR amplification. If there is a mismatch at the primer junction, it will be discriminated against by the thermostable ligase and the primers will not be ligated. The absence of the ligated product therefore indicates at a minimum a single basepair change in the target sequence (Barany, 199ab). Ligase detection reaction (LDR) is a method similar to LCR. In LDR one pair of adjacent primers, rather than two pairs as in LCR, is used therefore only a linear amplification process can be achieved.

Birkenmeyer and Mushahwar (1991) described a „gapped LCR", another ligase mediated technique which has also been termed pLCR (Barany 1991b). This technique uses four oligonucleotides primers with the two primers of one pair being separated by a gap of one or

Table 1. Current applications of LCR

Target	Format	Author
Genetic diseases		
ß-sickle cell hemoglobinemia	LCR, isotopic	Barany (1991a)
ß-sickle cell hemoglobinemia	LCR, fluorescent	Winn-Deen and Iovannisci (1991)
Cystic Fibrosis	LCR and pLCR, isotopic	Fang et al (1992)
Leber's hereditary Optic Neuropathy	PCR-LCR, nonisotopic	Zebala and Barany (1993)
Hyperkalemic periodic paralysis	PCR-LCR, fluorescent	Feero et al (1993)
Bacteria		
Borrelia burgdorferi	LCR, nonisotopic	Hu et al (1991)
Listeria monocytogenes	PCR-LCR, nonisotopic	Wiedmann et al (1992)
Neisseria gonorrhoeae	pLCR, nonisotopic	Birkenmeyer and Armstrong (1992)
Erwinia stewartii	PCR-LCR, nonisotopic	Wilson et al (1994)
Mycobacterium tuberculosis	LCR, fluorescent	Iovannisci and Winn-Deen (1993)
Chlamydia trachomatis	pLCR, isotopic	Dille et al (1993)
Viruses		
Human papillomavirus	LCR, nonisotopic	Bond et al (1990)
Herpes simplex virus	LCR, nonisotopic	Rinehardt e al (1990)
HIV DNA	LCR, nonisotopic	Carrino and Laffler (1990)
Feline immunodeficiency virus (FIV)	PCR-LCR, isotopic	Wiedmann and Batt (1993)
Other Targets		
Ha-*ras* protooncogene	LCR, nonisotopic	Kälin et al (1992)
G-6-PD	LCR, isotopic	Prechal et al (1993)

more differentiating bases which are specific for the target DNA. By including the missing nucleotides together with a thermostable polymerase in the reaction mix, first a gap filling reaction must be performed in the presence of the matching target and then the remaining nick can then be sealed by the ligase.

LCR assays have been developed for the detection of genetic diseases as well as the detection of bacteria and viruses. An overview over the current applications of LCR is shown in Table 1.

PERFORMANCE OF LCR REACTIONS

Accurate results from LCR assays depend on a variety of different factors including primer design and reaction conditions. In the following two sections we will review a few of the most important factors which need to be considered in the development of an LCR assay.

Design of LCR primers

A minimal amount of target independent ligation is observed when LCR primers with a single basepair overhang rather than blunt ends are used. The T_M of all four primers of one set of LCR primers should be within a narrow temperature range, ideally with an absolute T_M of $70°C\pm2°C$. Furthermore, the primers should be designed so that one primer does not contain sequences which will allow it to serve as a bridging template for two other primers and therefore lead to target-independent ligation. Adding non-complementary tails of two nucleotides or longer to the non-adjacent 3' ends of the primers will prevent ligation of these 3' ends. Depending on the nucleotides involved, different amounts of ligation product are to be expected with a mismatched target (Barany 1991a). Expected amounts of false ligation for certain mismatches are shown in Table 2. These data can be used for designing primers with the lowest possible rate of mismatches when a choice between different target sequences is permissible.

Table 2. Noise to signal ratio for certain mismatches in the LCR; * calculated as amount of product with mismatched primers divided by the amount of product with complementary primers (for details see Barany, 1991a).

Oligonucleotide base-target base	Noise to signal ratio, %*
A-A, T-T	1.1
T-T, A-A	<0.2
G-T, C-A	1.3
G-A, C-T	<0.2

LCR conditions

Standard conditions for a 50 µl LCR are as follows: One set of primers (between 25 and 200 fmol of each primer) are incubated in the presence of target DNA in the reaction buffer (50 mM Tris-HCl [pH 7.6], 100 mM KCl, 10 mM MgCl$_2$, 1 mM EDTA, 10 mM dithiothreitol,

1 mM NAD$^+$, 20 µg salmon sperm DNA) with 75 nick closing units of *T. aquaticus* DNA ligase (purified as described previously [Barany and Gelfand, 1991]) and overlaid with 50 µl of mineral oil. The inclusion of 0.01 % Triton X-100 in the reaction buffer gives a higher ligation rate, but also leads to a slight increase of ligation with a mismatched target (Winn-Deen and Iovannisci, 1991; Wiedmann et al., 1993). Reaction cycles are usually 1 min at 94°C for denaturation, followed by 4 min at 65°C (when the T_M of the primers is approximately 70°C). This cycling pattern is repeated between 10 and 30 times and the number of cycles has to be optimized for each LCR assay.

DETECTION METHODS FOR LCR PRODUCTS

Detection of the LCR product, i.e. the two ligated LCR primers was initially achieved by using radioactive labeled primers. Herein the separation of LCR products and primers was achieved by denaturing gel electrophoresis, so that the LCR product can be recognized by subsequent autoradiography. The level of sensitivity reached with this detection method is about 200 target DNA molecules (Barany, 1991a). Winn-Deen and Iovannisci (1991) described a nonisotopic detection method using fluorescent labeled primers. Specific detection of the LCR product can then be accomplished using a fluorescent DNA sequencer in conjunction with a GENESCANNER™ (Applied Biosystems). One of the advantages of this method is that it is relatively easy to quantitate the amount of the LCR products. Furthermore, this detection system allows the performance and analysis of multiplex LCR reactions when the different LCR primers are labeled with distinguishable fluorescent tags (Feero et al., 1993). The main disadvantage of this method, however, is the requirement of very expensive and sophisticated equipment. Therefore, the use of any of the two above described methods is not facile in a routine diagnosis laboratory.

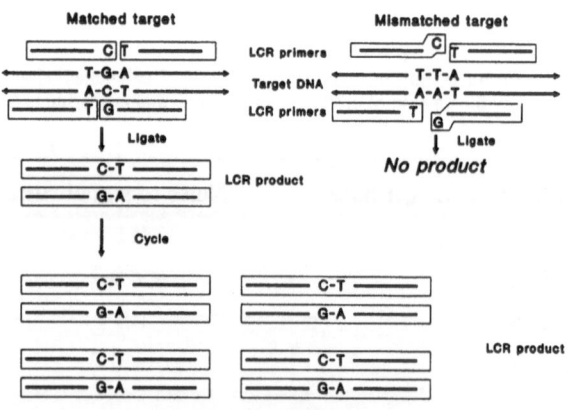

Figure 1. Schematic of LCR

Recently, more convenient methods for the detection of LCR products in microtitreplates have been developed (Wiedmann et al 1993; Winn-Deen et al 1993). For this purpose one LCR primer of a pair has been labeled with a biotin-"hook" at the 5' end while the other primer was labeled with a nonisotopic reporter at the 3' end. Reporter groups tested so far include fluorescein and digoxigenin. Methods utilizing a digoxigenin reporter molecule rely on the detection of this molecule with an enzyme-enhanced system. Anti-digoxigenin-antibodies, coupled to alkaline phosphatase (AP), are used to detect the digoxigenin. Subsequent detection of the AP can be achieved using chromogenic, fluorogenic or luminogenic substrates. Winn-Deen et al (1993) reported that a detection method using the luminogenic substrate Lumiphos™530 gave the highest sensitivity in this assay, which is only about 10 fold lower than detection methods using radioisotopes or fluorescent tagged LCR primers in conjunction with a fluorescent DNA sequencer. Another nonisotopic detection method for LCR products has been reported by Zebala and Barany (1993). They utilized primer pairs in which one primer is labeled with a poly(dA) tail at the 5' end while the other primer is tagged with biotin at the 3' end. After sufficient product is formed, the ligated product can be captured from the solution via hybridization of their poly(dA) tails with poly(dT) coated paramagnetic iron beads and subsequent magnetic separation. Only the captured LCR products will carry a 5' coupled biotin molecule, which can be detected with streptavidin-AP conjugate and subsequent addition of a chromogenic substrate.

Recent experiments in our laboratory showed the feasibility of detecting LCR products using streptavidin coated Biochips™ and a laser detection system (Tsay et al 1991). In this format Biochips™ are coated with streptavidin and LCR products, tagged with digoxigenin and biotin as described above, are hybridized to these chips. After denaturing washing steps and probing with a gold-coupled anti-digoxigenin antibody, a positive LCR result is characterized by an enhanced laser diffraction pattern (Piani et al 1993).

For the detection of the products from a pLCR, two different methods have been described. For an isotopic detection format radioactive labeled nucleotides, which are necessary to fill in the gap, are included in the reaction mix, so that the pLCR products can be detected by autoradiography after gel electrophoresis (Dille et al 1993). Nonisotopic detection of pLCR or gap LCR can be achieved by using pairs of pLCR primers labeled with biotin or fluorescein, respectively. Ligated oligonucleotides are captured on antifluorescein coated microparticles and detected with an antibiotin-AP conjugate. AP activity is subsequently detected with the fluorogenic substrate methylumbelliferone-phosphate (Birkenmeyer and Armstrong 1992).

DETECTION OF GENETIC DISEASES USING LCR

In one of the first published reports describing LCR, this technique was used to discriminate between normal β^A- and sickle β^S-globin genotypes in humans using either an isotopic detection method (Barany 1991a) or fluorescent-labeled LCR primers (Winn-Deen and Iovannisci 1991). For this purpose two sets of LCR primers were used, one specific for the normal allele and the other specific for the mutation. These two primers were applied in two separate LCR reactions and the LCR products analyzed separately.

Recently LCR has been exploited for the detection of other mutations which are the responsible for genetic disorders in humans. Examples include cystic fibrosis (Fang et al 1992), Leber's hereditary neuropathy (Zebala and Barany 1993) and hyperkalemic periodic paralysis (Ferro et al 1993). Further applications of the LCR in this area are expected to arise as progress in the human genome project results in the identification of the molecular origin of additional genetic diseases.

DETECTION OF BACTERIAL PATHOGENS USING LCR

Given the potential of LCR, attempts have been reported using this technique for the identification and detection of bacterial pathogens. Detection systems for bacteria based on the PCR usually depend on the availability of well characterized and specific target genes. This strategy is easily applied to extensively documented bacterial pathogens, where the sequence of one or more genes, associated with pathogenicity or virulence, is known. However, for many plant and animal pathogens as well as non-pathogenic bacteria from environmental sources, there is often not enough sequence information available to design species-specific PCR primers. The 16S rDNA, encoding part of the ribosomal RNA, consists of both highly conserved and variable regions, the latter usually contains at least a single basepair difference that is species-specific. A general method for PCR amplification and sequencing of this gene has been described by Weisburg et al (1991). Our group utilized these techniques to initially sequence the 16S rDNA gene of different isolates of the human pathogen *L. monocytogenes* and the closely related non-pathogenic bacterium *L. innocua* (Czajka et al 1993). We identified seven, consistent single basepair differences specific for *L. monocytogenes*, which were used to design LCR primers for the identification of this bacterium. To improve the sensitivity of this LCR, we further employed a set of flanking PCR primers to amplify the segment containing the specific single basepair differences prior to the LCR (Wiedmann et al 1992). This PCR-coupled LCR was shown to be highly specific for *L. monocytogenes* and was able to detect at a minimum, 100 CFU *L. monocytogenes*.

Our group used the same approach to develop an LCR based detection method for the plant pathogen *Erwinia stewartii* (Wilson et al 1993). After sequencing parts of the 16S rDNA gene of *E. stewartii* and the closely related saprophyte *E. herbicola*, we identified *E. stewartii*-specific single basepair differences. These were again used to design LCR primers for a PCR-coupled LCR, which proved to be specific for *E. stewartii*.

The development of these two PCR-coupled LCR for the detection of *L. monocytogenes* and *E. stewartii* suggests that this system is generally applicable for the development of a sensitive detection assay for bacteria when little or no prior genetic information is available.

Another approach for the detection of bacterial pathogens using LCR was tested by Iovannisci and Winn-Deen (1993). This group utilized LCR to detect *Mycobacterium tuberculosis* DNA, based on a specific insertion sequence, IS6110. Using fluorescent labeled primers and a fluorescent DNA sequencer they were able to detect as few as 200 copies of the target molecule even in the presence of unrelated DNA.

Assays for the detection of the bacterial pathogens *Neisseria gonorrhoeae* and *Chlamydia trachomatis* using pLCR have also been described (Birkenmeyer and Armstrong,1992; Dille et al 1993). These assays are based upon two basepair differences between the target bacterium and closely related non-pathogenic bacteria. The sensitivity of these assays is 1.1 *N. gonorrhoeae* cells using the nonisotopic detection method and 3 *C. trachomatis* elementary bodies using the isotopic detection method with electrophoretic separation of the products from unligated primers as described above. Detection of *N. gonorrhoeae* was achieved by using pLCR probes targeting sequences in the gene for the cell surface opacity (Opa) protein or in the gene for the pilin proteins (Birkenmeyer and Armstrong 1992). The targeted sequences show only two basepair differences between *N. gonorrhoeae* and the closely related *N. meningitidis*, which is sufficient for clear differentiation by pLCR. For the specific detection of *C. trachomatis*, pLCR primers were used that recognized species-specific sequences either in the gene for the major outer membrane protein or on a cryptic plasmid (Dille et al. 1993).

DETECTION OF VIRUSES USING LCR

Only a few preliminary reports on the use of LCR for the identification and/or detection of viruses have been published (see Table 1). Detection of HIV was performed using LCR oligonucleotides targeting part of the gag region of HIV-1. The sensitivity of this assay is between 5 and 10 HIV-1 molecules, which compares to the level of sensitivity reached by PCR (Carrino and Laffler 1990). LCR technology was also applied for the detection of herpes simplex virus and allows rapid detection of this virus as compared to traditional detection methods using cell culture techniques (Rinehardt et al 1990). Our group is currently investigating the use of LCR for the specific identification of certain strains of the Feline Immunodeficiency Virus (FIV) based on single basepair differences. These experiments will also show whether thermostable Taq ligase is able to work on hybrids between DNA primers and RNA targets. This would prove to be a big advantage for the direct detection of RNA viruses as well as mRNA targets. Application of the LCR in virological diagnostics might furthermore prove useful for the identification of single basepair differences between different viral strains, which may be correlated to pathogenic behavior.

FUTURE PERSPECTIVES

With the continuing emergence of sequence data for the human genome as well as the genomes of other species (e.g. bovine) the potential of LCR to detect genetic diseases which result from single basepair mutations is immense. The inherent advantages of LCR are its compatibility with primary target amplification using PCR or 3SR and its high specificity. Furthermore, LCR is highly amenable to automation since the LCR product consists of two covalently joined primers which can be easily detected using different enzyme-linked or direct fluorescent labels. The formatting of multiplex LCR assays will further improve screening samples for an array of different potential single basepair changes in a single tube format. An automated, multiplex LCR or PCR-coupled LCR assay has a variety of potential applications such as: (1) screening of large populations for monogenic disease polymorphisms; (2) determine HLA haplotypes in tissue typing, e.g. for transplantation; (3) screening for multiple bacterial species after a generic PCR amplification of 16S rDNA sequences.

In clinical diagnosis of pathogenic bacteria and viruses the specificity of LCR will be important in many applications. The specific detection of single basepair differences in bacterial pathogens may be valuable with respect to antibiotic resistance which can be based on point mutations, e.g in some cases of macrolide resistance (Gauthier et al 1988) or on transformational exchange which leads to diagnostic single basepair differences between sensitive and resistant strains, e.g. in *Neisseria meningitidis* (Radström et al 1992). In viral pathogens, the identification of subpopulations with genetic differences which may affect the host range, virulence characteristics and drug resistance may be insightful.

Furthermore, the application of LCR and PCR-coupled LCR assays for the detection of specific bacteria based on at least single basepair differences in the 16S rDNA gene has great potential. As outlined above this method circumvents the need to identify species specific genes, which could serve as a target for PCR or other nucleic acid based assays. With emerging interest in yet poorly characterized bacteria, e.g. soil bacteria, this method should have a great perspective as a detection system.

ACKNOWLEDGEMENTS

This work was supported by the Northeast Dairy Foods Research Center, a grant from the Cornell Biotechnology Program which is sponsored by the New York State Science and Technology Foundation, a consortium of industries, the U.S. Army Research Office and the National Science Foundation to C.B. and the National Institutes of Health (GM 41337-03) to F.B. M.W. was supported by a stipend of the Gottlieb Daimler- and Carl Benz-Stiftung (2.92.04).

REFERENCES

Barany F. The ligase chain reaction in a PCR world. PCR Methods Applicat 1991a, 1:5-16.

Barany F. Genetic disease detection and DNA amplification using cloned thermostable ligase. Proc Natl Acad Sci USA 1991b, 88:189-193.

Barany F and Gelfand D. Cloning, overexpression and nucleotide sequence of a thermostable DNA ligase-encoding gene. Gene 1991, 109:1-11.

Birkenmeyer LG and Mushahwar IK. Mini-Review: DNA probe amplification methods. J Virol Methods 1991, 35:117-126.

Birkenmeyer L and Armstrong SA. Preliminary evaluation of the ligase chain reaction for specific detection of Neisseria gonorrhoeae. J Clin Microbiol 1992, 30:3089-3094.

Bond S, Carrino J, Hampl H, Rinehardt L and Laffler T. New methods of detection of HDV. 1990, In Serono Symposia (ed. Monsonego J). Raven Press, Paris.

Carrino JJ and Laffler TG. Detection of HIV-DNA sequences using the ligase chain reaction (LCR). Clin Chem 1991, 37:1059.

Czajka J, Bsat N, Piani M, Russ W, Sultana K, Wiedmann M, Whitaker R and Batt CA. Differentiation of Listeria monocytogenes and Listeria innocua by 16S rDNA genes and intraspecies discrimination of Listeria monocytogenes strains by random amplified polymorphic DNA polymorphisms. Appl Environ Microbiol 1993, 59:304-308.

Dille BJ, Butzen CC and Birkenmeyer LG. Amplification of Chlamydia trachomatis DNA by ligase chain reaction. J Clin Microbiol 1993, 31:729-731.

Fang P, Jou C, Bouma S and Beaudet A. Detection of cystic fibrosis mutations using the ligase chain reaction. Am J Hum Genet 1992, A214.

Feero WG, Wang J, Barany F, Zhou J, Todorovic SM, Conwit R, Galloway G, Hausmanowa-Petrusewicz I, Fidzanska A, Arahata K, Wessel HB, Wadelius C, Marks HG, Hartlage P, Hayakawa H and Hoffman EP. Hyperkalemic periodic paralysis: rapid molecular diagnosis and relationship of genotype to phenotype in 12 families. Neurology 1993 (in press).

Gauthier A, Turmel M and Lemieux C. Mapping of chloroplast mutations conferring resistance to antibiotics in chlamydomonas: evidence for a novel site of streptomycin resistance in the small subunit ribosomal RNA. Mol Gen Genet 1988, 214:192-197.

Huh, Elmorek, Facey Iand Jenderzak D. Detection of *Borrelia burgdorferi* by ligase chain reaction. Abstr Gen Meet Am Soc Microbiol 1991, 79.

Iovannisci DM and Winn-Deen ES. Ligation amplification and fluorescence detection of *Mycobacterium tuberculosis* DNA. Mol Cell Probes 1993, 7:35-43.

Kälin I, Shephard S and Candrian U. Evaluation of the ligase chain reaction (LCR) for the detection of point mutations. Mutation Res 1992, 283:119-123.

Landegren U, Kaiser R, Sanders J and Hood L. A ligase-mediated gene detection method. Science 1988, 241:1077-1080.

Nickerson DA, Kaiser R, Lappin S, Stewart J, Hood L and Landegren U. Automated DNA diagnostics using an ELISA based oligonucleotide ligation assay. Proc Natl Acad Sci USA 1990, 87:8923-8927.

Piani M, Rocco R and Batt CA. Unpublished Results. 1993.

Prchal JT, Guan YL, Prchal JF and Barany F. Transcriptional analysis of the active X-chromosome in normal and clonal hematopoiesis. Blood 1993, 81:269-271.

Radström, P, Fermer, C, Kristiansen, B-E, Jenkins, A, Sköld, O and Swedeberg, G. Transformational exchanges in the dihydropteroate synthase gene of *Neisseria meningitidis*: a novel mechanism for acquisition of sulfonamide resistance. J Bacteriol 1992, 174: 6386-6393.

Rinehardt L, Hampl H and Laffler TG. Ultrasensitive non-radioactive detection of herpes simplex virus by LCR, the ligase chain reaction. 20th Annual Meeting of the Keystone Symposia on molecular and cellular biology 1991, p. 101.

Tsay YG, Lin CI, Lee J, Gustafson EK, Appelqvist R, Magginetti P, Norton R, Teng N and Charlton D. Optical biosensor assay. Clin Chem 1991, 37:1502-1505.

Weisburg WG, Barns SM, Pelletier DA and Lane DJ. 16S ribosomal DNA amplification for phylogenetic study. J. Bacteriol. 1991, 173:697-703.

Wiedmann M, Czajka J, Barany F and Batt CA. Discrimination of *Listeria monocytogenes* from other *Listeria* species by ligase chain reaction. Appl Environ Mirobiol 1992, 58:3443-3447.

Wiedmann M, Barany F and Batt CA. Detection of *Listeria monocytogenes* in surface swab samples using a nonisotopic polymerase chain reaction (PCR)-coupled ligase chain reaction (LCR) assay. Appl Environ Micorbiol 1993, 59: 2143-2145

Wiedmann M and Batt CA. Unpublished Results. 1993.

Wilson WJ, Wiedmann M, Dillard HR and Batt CA. Identification of *Erwinia stewartii* by a ligase chain reaction assay. Appl Environ Microbiol 1994, 60: 278-284

Winn-Deen ES and Iovannisci DM. Sensitive fluorescence method for detecting DNA ligation amplification products. Clin Chem 1991, 37:1522-1523.

Winn-Deen ES, Batt CA and Wiedmann M. Non-radioactive detection of *Mycobacterium tuberculosis* LCR products in a microtitreplate format. Mol Cell Probes. 1993 (accepted).

Zebala JA and Barany F. Detection of Leber's hereditary optic neuropathy by non radioactive-LCR. *In* Innis MA, Gelfand DH, Sninsky JJ and White TJ (ed.), PCR Protocols: A guide to Methods and Applications., 2nd ed. Academic Press Inc. San Diego 1993 (accepted).

QUALITATIVE AND QUANTITATIVE DETECTION OF NUCLEIC ACIDS OF INFECTIOUS AGENTS BY NASBA

B. van Gemen, T. Kievits and P.F. Lens

Organon Teknika, Boxtel, The Netherlands

INTRODUCTION

Since the advent of the polymerase chain reaction (PCR; Saiki 1988; Mullis & Faloona 1987), a number of other nucleic acid amplification techniques have been developed. One of the most important and well developed of these new technologies is the Nucleic Acid Sequence Based Amplification (NASBA™ method (Kievits et al 1991). NASBA™ utilizes the coordinated activities of AMV reverse transcriptase (RT), RNase H, and T7 RNA polymerase to amplify a specific nucleic acid target. The specificity of the reaction is determined by a pair of oligonucleotide primers, which are specific for the sequence of interest. One of these primers (designated P1) is synthesized so as to include the promoter for T7 RNA polymerase as a 5' overhang. The reaction is conducted at constant temperature (41°C) and produces a single stranded RNA product which represents a 10^6 - 10^9 amplification of the original target sequence (figure 1). Although capable of amplifying both DNA and RNA target sequences, NASBA™ is most suitable for the amplification of RNA. Thus, NASBA™ has become an extremely powerful technique for the detection and quantification of retroviruses (particularly HIV-1).

Importantly, NASBA™ has also been developed for the detection of a number of other viruses (eg. HPV, CMV, and HCV). However, its use has not been limited to the detection of viruses. The technique has also been used to screen food for bacterial pathogens, Further, NASBA™ has been utilized to detect specific growth factor transcripts. Clearly, the power and versatility of the technique has allowed NASBA™ to be developed into an important mean for molecular characterization of nucleic acids.

There are various means of detecting nucleic acid amplification products. Acrylamide or agarose gel diagnosis of amplification reactions will indicate if a product of the appropriate

Methods in DNA Amplification, Edited by
A. Rolfs *et al.*, Plenum Press, New York, 1994

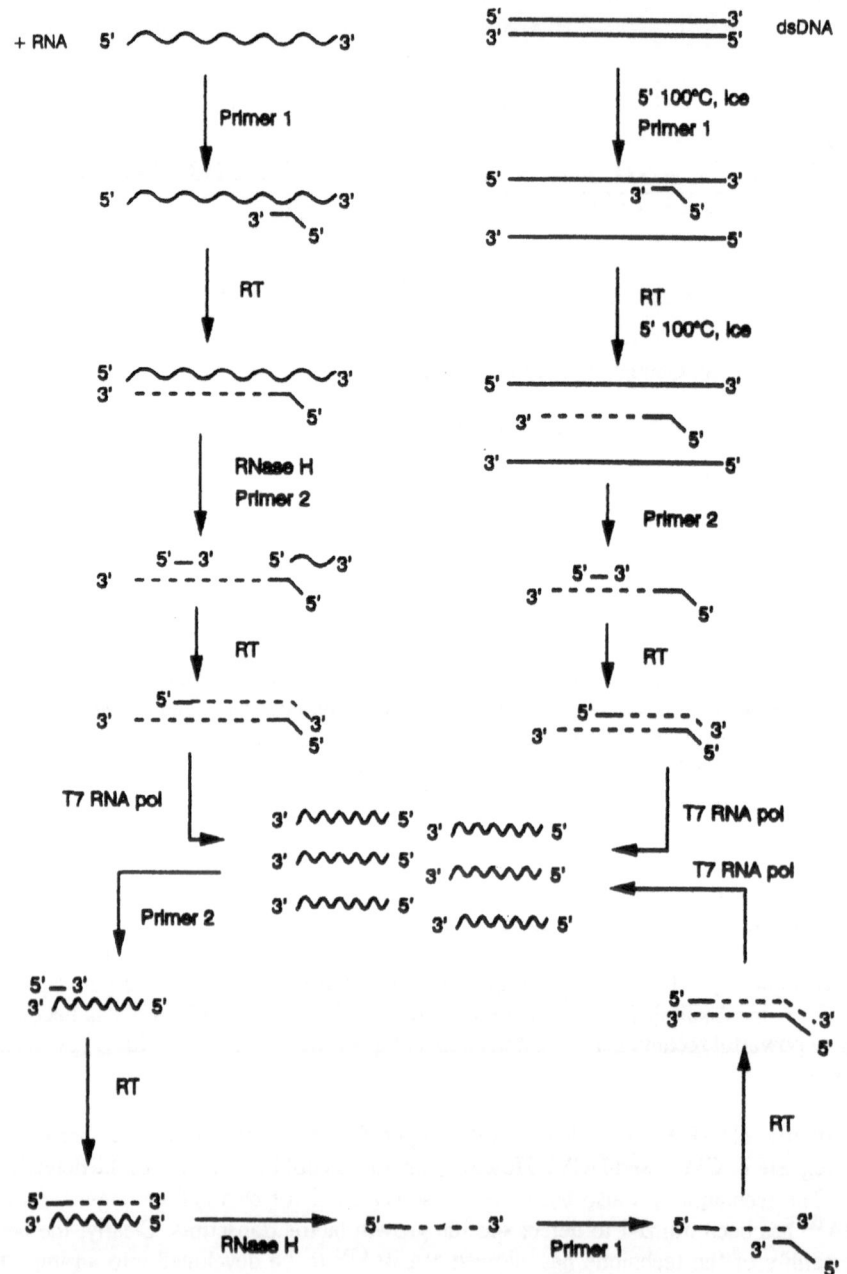

Figure 1. Schematic presentation of the NASBA™ method applied to ssRNA and dsDNA. The method is based on extension of Primer 1 (T7-containing) by AMV-reverse transcriptase, degradation of the RNA strand by RNase H in a RNA/DNA template (or heat denaturation for dsDNA), synthesis of second DNA strand by AMV-RT and RNA synthesis by T7 RNA polymerase. With RNA synthesis the system enters the cyclic phase, that is based on the principles described above. R.T. = AMV reverse transcriptase. Solid lines represent target DNA, broken lines newly synthesized DNA and wavy lines represent RNA. For further explanation see text.

size was produced, however this will not prove that the product was truly obtained from the desired target sequence. Typically, amplification products are small and accurate sizing is difficult. Therefore, the conventional means of diagnosing an amplification reaction product has been through hybridization analysis. The hybridization procedure requires an oligonucleotide probe that is specific for a portion of the target sequence which is located between (but does not include) the primer sequences. PCR products are routinely diagnosed by Southern transfer of the amplification product to a nylon or nitrocellulose membrane, followed by hybridization with a radiolabeled oligonucleotide probe. Alternatively, PCR products can be diagnosed by a liquid hybridization procedure which involves incubating the amplified product with the radiolabeled probe at an appropriate temperature. After sufficient incubation, the product-probe mixture is resolved on an acrylamide gel. Product which is hybridized to the probe will migrate to a specific position in the gel, whereas non-hybridized probe will advance further into the gel. Autoradiography is used to visualize the results obtained from either of these procedures.

Amplified products obtained by NASBA™ can be diagnosed with the same hybridization strategies used to diagnosis PCR products. However, since the product of the NASBA™ reaction is single stranded, it can be diagnosed by hybridization much more easily than a PCR product.

Importantly, non-radioactive methods for the detection of NASBA amplified target sequences have been developed. Oligonucleotide probes conjugated to horse radish peroxidase (HRP) allow for liquid hybridization followed by a gel-retardation assay and staining of the HRP in the gel. This so called enzyem linked gel assay (ELGA) has a number of advanges over the use of radiolabelled probes on Southern blots.

The most obvious is the fact that the procedure is nonisotopic. Further, liquid hybridization is more rapid and much simpler than Southern transfer and subsequent hybridization. Since the color reaction mediated by HRP is very rapid (typically less than two minutes), results are obtained much quicker than with autoradiography. NASBA™ has currently been developed to the point where this nonisotopic detection method could be used to detect target sequences at multiple positions over the HIV-1 genome.

DETECTION OF HIV-1 BY PCR AND NASBA

Recently, NASBA™ has been used to screen whole blood and blood fractions of HIV-1 infected individuals (Bruisten et al 1993) in order to determine the presence of HIV-1. The same set of samples were analyzed by NASBA™, DNA PCR and RT-PCR of RNA. In this study, nucleic acids were extracted from whole blood, mononuclear cells, plateletes, and plasma using a guanidinium thiocyanate-silica extraction method (Boom er al 1991). Donors were grouped according to their level of CD4+ cells (group A, >0.5 x 10^9/L; group B, 0.20 < < 0.50 x 10^9/L; group C, < 0.20 x 10^9/L). The results of the study are summarized in Table 1.

Interestingly, it was demonstrated that RT-PCR of RNA was more sensitive than DNA PCR for the detection of HIV-1 in whole blood. Importantly, NASBA™ was shown to be as sensitive. DNA PCR could not be used to screen plasma (which lacks cells) or platelets (which are anuclear). When RT-PCR and NASBA™ are compared in the screening of these two blood fractions, NASBA™ is as sensitive, or more sensitive, than RT-PCR. As the results of this study indicate, NASBA™ is as effective as PCR for the screening of whole blood or blood fractions for HIV-1. In view of the many technical advantages of NASBA™ over PCR (which were outlined earlier), it is clear how this technology would greatly benefit research efforts requiring HIV-1 detection.

Table 1. Amplification results startified by CD4+ level. [a]Group A, >0.50; group B, >0.20 <0.50; group C, <0.20 (CD4+ cells x 109/liter). [b]PBMCs, Peripheral blood mononuclear cells. [c]Number of amplification-postive nucleic acid samples per total number of samples tested (percentages in parentheses). [d]The HIV-1 DNA PCR-positive platelet samples are also PCR postive for ß-goblin PCR, indicating contamination of these fractions with white blood cells

Amplification procedure	CD4+ group[a]	Whole blood	Plasma	PBMC[b]	Platelets
DNA PCR	A	8/18 (44)[c]	0/18 (0)	17/18 (94)	1/18[d] (6)
	B	17/23 (74)	0/23 (0)	23/23 (100)	0/19 (0)
	C	24/29 (83)	0/29 (0)	29/29 (100)	2/20[d] (10)
	Total	49/70 (70)	0/70 (0)	69/70 (99)	3/57[d] (5)
RT-PCR	A	12/18 (67)	9/18 (50)	17/18 (94)	6/18 (33)
	B	19/23 (83)	20/23 (87)	22/23 (96)	6/21 (29)
	C	28/29 (97)	27/29 (93)	29/29 (100)	17/24 (71)
	Total	59/70 (84)	56/70 (80)	68/70 (97)	29/63 (46)
NASBA	A	9/18 (50)	10/18 (56)	14/18 (78)	5/18 (28)
	B	22/23 (96)	21/23 (91)	22.23 (96)	13/21 (62)
	C	28/29 (97)	29/29 (100)	28/29 (97)	22/24 (92)
	Total	59/70 (84)	60/70 (86)	64/70 (91)	40/63 (64)

USE OF NASBA™ FOR THE DETECTION OF POINT MUTATIONS

A number of specific mutations within the HIV-1 pol gene have been identified which are critical for the establishment of resistance to AZT (Larder et al 1989). Although these mutations were originally identified by DNA sequence analysis, that approach is not practical for the screening of large numbers of AZT resistant (AZT[R]) isolates. NASBA™, coupled with specific oligonucleotide probe hybridization analysis, has been used to screen HIV-1 isolates for the characteristic mutations which typify AZT resistant strains. Oligonucleotide probes, specific for wild type and AZT[R] mutant sequences at codons 70 ans 215 of the pol gene, were synthesized. The single stranded RNA product obtained with the NASBA™ reaction was easily diagnosed by differential hybridization with the wild type and mutant-specific probes. Thus, NASBA™ provides a powerful means of rapidly screening for relevant point mutations. This application has important implications for the monitoring of drug therapy of AIDS patients.

QUANTIFICATION OF HIV-1

Nucleic acid amplification has a number of important advantages over other methods of virus quantification. For example, assays measuring the level of p24 gag protein or reverse transcriptase activity are dependent upon the specific expression of viral genes. Quantification of antibody levels in the host is also dependent upon expression of viral antigens, as well as an immune response by the host. The use of a syncitia formation assay is limited by the functional capability of the particular isolate being tested. None of these limitations apply to virus quantitation by a nucleic acid amplification method.

Quantitative PCR has been developed for determining virus load in HIV-1 infected individuals (Ho et al 1989; Pantaleo et al 1991). These studies utilized quantitative PCR to determine the level of integrated provirus in the peripheral blood, or specific tissues, of infected individuals. In fact, these studies served to demonstrate that nucleic acid amplification is a far more sensitive and accurate means of establishing virus load than any of the methods which are dependent upon virus activation or gene expression. It is also possible to use quantitative RT-PCR (Zack et al 1990) to assess the level of virus genome or transcript.

The quantitative basis of most of these techniques involves the coamplification of a control molecule (e.g. plasmid with a cloned target sequence), or the extracted material from a control cell line such as ACH2, which is known to harbor a single copy of the HIV-1 genome (Schnittman et al 1989). The quantitative data obtained from these control amplifications is then used as a standard for the quantitation of unknown samples. For example, if hybridization of a radiolabeled probe to amplified material is the means of detecting product, a standard curve can be generated by subjecting control quantities of target material to densitometric analysis. Comparisons can then be made between the signal intensity of unknown samples, with the signals obtained from the control samples.

Another strategy has been used to adapt NASBA™ technology into a quantitative method. A number of important modifications have been developed so as to establish NASBA™ as a more accurate and simple means of HIV-1 quantification, relative to other amplification techniques. The critical component to quantitative NASBA™ is the proper development of a control target sequence. A 1.4 kb fragment from the HIV-1 genome was cloned into the pGEM plasmid. The cloned fragment includes a portion of the gag region for which a set of NASBA™ primers and a detection probe have been developed.

In addition to this wild type HIV-1 pGEM construct, a second, control pGEM construct was developed. It is identical to the wild type construct, however a 20 base region within the the area amplified by the NASBA™ primers has been mutagenized. This 20 base region was removed,and its sequence was randomized in such a way that the A, C, G and T frequencies were maintained. Thus, RNA produced from this control construct (designated Q-RNA) could be amplified with the same set of NASBA™ primers that would amplify wild type RNA, and the product would be the exact same size. Since the wild-type and Q-RNAs would produce the same size fragment upon amplification with the same primer set, the kinetics of amplification for each molecule would be identical. However, the Q-RNA could be distinghuished from the wild type RNA with a probe specific for the 20 base randomized sequence. Since the A, C, G and T frequencies of the Q-RNA probe target sequence are the same as those of the wild type probe target, the required wash stringencies of both probes would be the same.

Once these wild type and Q-RNAs were obtained from the corresponding pGEM plamids, they could be used in quantitative NASBA™. Figure 2 summarizes the strategy of the assay. In this example a control quantity of wild type RNA will be assayed. Six NASBA™ reactions, each containing 10^4 copies of wild type RNA, are set up. Five of these reactions are then spiked individually with 10^2, 10^3, 10^4, 10^5 and 10^6 copies of control Q-RNA. The sixth receives no Q-RNA and will be an amplification of wild type material only. This amplification series will compete for the reagents present in the reaction mix. Since the A, C, G, and T content of both templates is identical, and amplification of both Q and wild type RNA produces the same size product, each will be equally competitive for reagents. Therefore, the degree to which these two species are amplified will be directly proportional to the starting quantity of each species present in the reaction. When the amount of wild type RNA in the reaction equals the amount of spiked Q-RNA, 50% of the product will be Q-RNA, and 50% will be wild type. At this concentration of Q-RNA, the amount of wild type product will be 1/2 the amount producted in the unspiked wild type reaction.

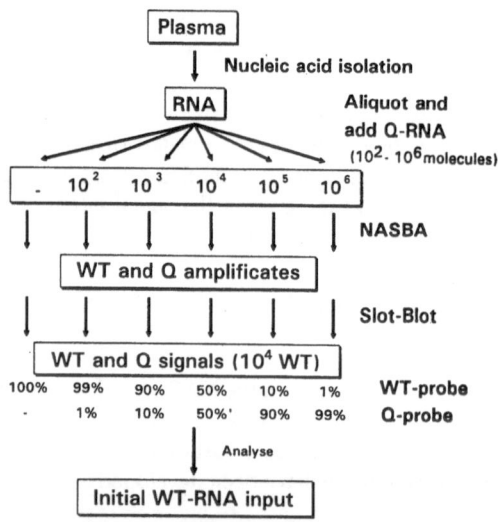

Figure 2. Schematic presentation of the experimental setup for a Quantitative NASBA (Q-NASBA). Relative signal intensity for both wild-type and Q probes is given for a model system with an initial input of 10^4 wild-type RNA molecules.

Since the Q-RNA product can be distinguished from the wild type RNA product by virtue of the unique 20 base sequence described earlier, it is possible to use hybridization analysis to determine the amount of each product type produced in a given reaction. After amplification as described above, the reaction products can be transferred in duplicate to a nylon membrane, by means of a vacuum slot blot apparatus. One set of transferred reaction products is hybridized with the wild type radiolabeled probe, the other with the radiolabeled Q-RNA-specific probe. Since both probes will bind with equal intensity, the blot washings can be done together at the same stringencies. Densitometric analysis of the resulting autoradiograph will indicate what quantity of Q-RNA is needed to reduce by 50% the wild type probe signal obtained by hybridization to the unspiked sample. At this concentration of Q-RNA, the signal intensity obtained with the Q-probe should also be reduced 50% compared to separately amplified Q-RNA. By scoring these intensities, the amount of wild type RNA can be accurately determined.

The above described strategy for quantitative NASBA™ has been coupled with a nonisotopic hybridization detection method. This approach utilized the capacity to readily synthesize oligonucleotide probes which have a biotin group at the 5' end. Such biotinylated oligonucleotides can then be linked to magnetic beads which have been coated with streptavidin. When the sequence of these bead conjugated oligonucleotides is specific for a region of amplified material (i.e. amplificate), the conjugated probe can serve to capture amplificate onto the bead. Next, the horse radish peroxidase conjugated oligonucleotide probe can be used to detect the captured amplificate. The color reaction could then be used to quantitate the amount of captured product.

This technique has a number of important attributes. First, it is nonisotopic. Unlike the hybridization technique described in the previous section. results can be obtained here with a color development. Further, by capturing amplified material onto a magnetic bead the entire detection proces is much faster than transfer to a nylon membrane and solid phase hybridization. Amplified products can be captured onto beads, washed, reacted with the detection probe, and color intensity determined in approximately three hours. The procedure is made simple by the capacity to hold the beads in place throughout the procedure with a magnetic rack. This allows for accurate quantitation analysis of the reaction products and lends the entire process well to automation. The ease with which the magnetic bead-based quantitative NASBA™ can be applied to various biological samples establishes the technology as indispensable to projects requiring rapid and accurate quantification of HIV-1.

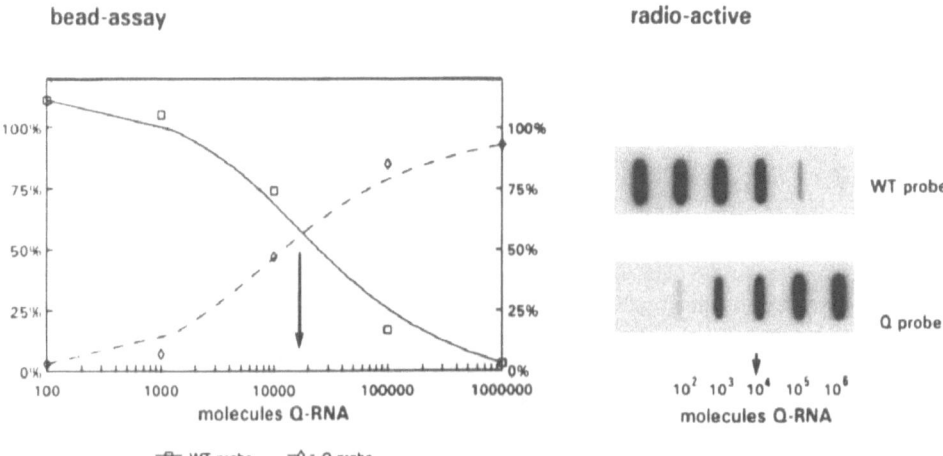

Figure 3. Quantification of a virus stock produced by HIV HXB3 infected H9 cells. Nucleic acid was extracted from 100 μl culture supernatant. Ten-thousand fold diluted nucleic acid was subjected to Q-NASBA. Amplificate was hybridized to HRP-labelled wild-type (WT) and Q-probes and the result is shown in the left panel. Amplificate was also slot-blotted and hybridized to ^{32}P-labelled wild-type (WT) and Q-probes (right panel). Both methods yield the same number of input RNA molecules.

The quantitative NASBA (Q-NASBA) coupled to both detection systems described using wild type and Q-RNA specific probes was tested using a model system with an input of 10^4 wild type RNA molecules. The result is depicted in figure 3. Both the radioactive and non-radioactive bead assay give accurate results.

USE OF NASBA FOR THE DETERMINATION OF VIRAL LOAD IN CLINICAL SAMPLES

The Q-NASBA procedure was used for analysis of the HIV-1 RNA titre in sequential plasma samples of an individual with a HIV-1 primary infection. The plasma samples were also examined for titres of HIV-1 p24 antigen.

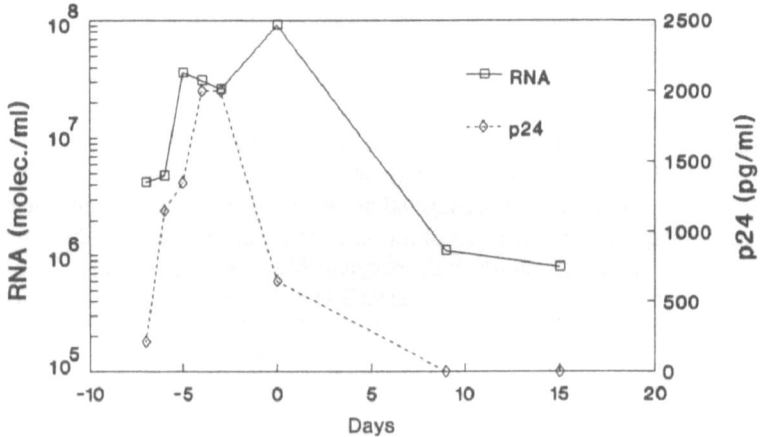

Figure 4. Graphic presentation of the HIV-1 RNA levels (squares) and p24 antigen levels (diamonds) of a patient with a HIV-1 primary infection. Seroconversion for antibodies against p24 and gp160 was on day 0.

Graphical display of the data (Figure 4) clearly shows coincidence of the RNA profile with the p24 antigen profile, albeit that the Q-NASBA procedure is a better indicator for presence of viral particles than the p24 antigen assay used in this study. With NASBA it is still possible to detect RNA (i.e. viral particles) after seroconversion while there is no detectable p24 antigen. Acidic treatment of the plasma samples in order to release HIV-1 p24 antigen from possible immune complexes (Nishanian et al, 1990) did not change the p24 antigen results. The peak of RNA level and p24 antigen level are on day 0 and days 4 and 3 respectively. The first positive results for RNA and p24 antigen were obtained 7 days before seroconverion. The decline in RNA titre and p24 antigen titre synchronizes with the elevation of antibodies against HIV-1.

CONCLUSIONS

Amplification technologies are powerfull mechanisms for the detection and quantitation of retroviruses. NASBA™ is a simple and reliable amplification technology which has been highly developed for the detection and quantitation of HIV-1. NASBA™ has a number of technical advantages over PCR; it is isothermal, eliminating the need for a thermal cycling unit; it is a single step RNA amplification method, eliminating the need for a preliminary cDNA synthesis step; and, it generates single stranded reaction products, which are readily diagnosed by hybridization analysis or can be subjected to sequence analysis directly. In terms of its specific application to HIV-1 detecting and quantification, NASBA™ has proven to be a very powerful technique. Although easily detected by hybridization analysis with radiolabeled probes, nonisotopic detection methods utilizing horse radish peroxidase conjugated oligonucleotide probes have been developed. NASBA™ has also been developed into a sensitive and accurate quantitaive assay which can be used to determine virus load in a wide array of biological specimens (e.g., blood, blood fractions, cells, tissue, faeces, urine, etc). The quantitative NASBA™ has been adapted into a colorimetric assay, using a 96 well microtiter plate format. This assay makes use of magnetic beads, coated with oligonucleotide capture probes, which are specific for amplified products. Thus, results obtained with this technique can be rapidly analyzed using a standard ELISA plate reader.

It is very clear that the highly developed state of NASBA™ technology for the detection and quantification of HIV-1, establishes it as one of the most powerful and essential molecular techniques available for the study of HIV and AIDS.

REFERENCES

Boom R., Sol C J A, Salimans M M M, Jansen C L, Wertheim - van Dillen P M E and van der Noordaa J. Rapid and simple method for purification of nucleic acids.J Clin Microbiol 1990, 28: 495 - 503.

Bruisten S, van Gemen B, Koppelman M, Rasch M, van Strijp D, Schukkink R, Beyer R, Weigel H, Lens P and Huisman H. Distribution of HIV-1 in different blood fractions of HIV-1 seropositive persons using two nucleic acid amplification assays. AIDS Hum Retrovir.In press.

Ho D.D., Moudgil T and Alam H. Quantification of human immunodeficiency virus type 1 in the blood of infected persons. New Eng. J. Med. 1989, 321: 1621 - 1625.

Kievits T, van Gemen B, van Strijp D, Schukkink R, Dircks M, Adriaanse H, Malek L, Sooknanan R and Lens P. NASBA™ isothermal enzymatic in vitro nucleic acid amplification optimized for the diagnosis of HIV-1 infection. J. Virol Meth. 1991, 35: 273 - 286.

Larder B.A., Darby G and Richman D D. HIV with reduced senstivity to zidovudine (AZT) isolated during prolonged therapy. Science 1989, 243: 1731.

Mullis K.B. and Faloona F A. Specific synthesis of DNA in vitro via a polymerase catalyzed chain reaction. Methods Enzymol 1987, 155: 335 - 350.

Nishanian P, Huskins K R, Stehn S, Detels R and . Fahey J L. A simple method for improved assay demonstrates that p24 antigen is presnt as immune complexes in most sera from HIV-infected individuals. J. Inf. Dis. 1990, 162: 21 - 28.

Pantaleo G., Graziosi C, Butini L, Pizzo PA, Schittman S M, Kotler D Pand Fanci A S. Lymphoid organs function as major reservoirs for human immunodeficiency virus. Proc. Natl. Acad. Sci. USA 1991, 88: 9838 - 9842.

Saiki R K, Gelfand D H, Stoffel S, Scharf S J, Higuchi R, Horn G T, Mullis K B and Ehrlich H A. Primer-dependent enzymatic amplification of DNA with a thermostable DNA polymerase. Science 1988, 239: 487 - 491.

Zack J.A., Arrigo S J, Weitsman S R, Go A S, Haislip A and Chen ISY. HIV-1 entry into quiescent primary lymphocytes: molecular analysis reveals a labile latent viral structure. Cell 1990, 61: 213 - 222.

APPLICATION OF NASBA TO THE DETECTION OF LISTERIA MONOCYTOGENES

J. Kleiber, C. Kaletta, M. Hartl, C. Kessler, P. Kirch and C. Majewski

Boehringer Mannheim GmbH, Penzberg, Germany

INTRODUCTION

Listeria monocytogenes is associated with meningoencephalitis, septicemia and abortion in humans (Gray et al 1966). Pregnant women, newborns, and immunocompromised individuals are particularly susceptible to infection; however, in some instances, apparently healthy individuals develop clinical disease after ingestion of food contaminated with this pathogen (Schlech et al 1983; Schwartz et al 1989). There have been at least five serious foodborne outbreaks of listeriosis reported in North America and Europe in recent years (Schwartz et al 1989).

As a consequence, there is an increased demand for quick and reliable methods for detecting the bacterium in food products. Although cultural methods provide the „gold standard" for detection of Listeria, these are labor intensive and time consuming. A more rapid method which specifically identifies *L. monocytogenes* is required. Amplification by PCR and detection of organism-specific nuleic acid sequences has facilitated the diagnosis of various bacteria (Innis et al 1990).

Recently an alternative amplification method, named "Nucleic acid sequence based amplification" (NASBA), based on the simultaneous activity of three enzymes has been described (Cangene Corp., European Patent Application EPO329822; Compton 1991; Kievits et al 1991). As an alternative to PCR amplification, NASBA offers users the ability to amplify fragments of DNA or RNA. For RNA this method yields an amplification of a billion fold (Compton 1991). Since the process is isothermal, the method is even more suitable for automation than PCR.

In this article, we describe the evaluation of NASBA with respect to its specificity, sensitivity and lowest detection limit for *L. monocytogenes* DNA.

Methods in DNA Amplification, Edited by
A. Rolfs *et al.*, Plenum Press, New York, 1994

PRINCIPLE OF NASBA FOR AMPLIFICATION OF DNA

A NASBA assay for DNA is divided into two different steps, PreNASBA and NASBA itself (Fig. 1). In Pre NASBA, double-stranded DNA is denatured and hybridized with primer 1, which is about 45 bases in length carrying a 5′-end with a promoter sequence that is recognized by T7 RNA polymerase and a 3′-end that is complementary to the target sequence. AMV reverse transcriptase extends the 3′-end of primer 1, thereby forming a cDNA of the template. To obtain a single stranded DNA intermediate compatible for NASBA a second heat denaturing step is performed.

A standard NASBA reaction mixture contains T7 RNA polymerase, RNase H, AMV reverse transcriptase, nucleoside triphosphates, primer 1 and 2 and appropriate buffer components. In a first step primer 2 anneals to the single-stranded cDNA of PreNASBA. AMV reverse transcriptase extends the 3′end of primer 2, rendering the promoter region double-stranded. T7 RNA polymerase transcribes RNA copies from the now transcriptionally-active promoter generating as many as 100 copies from each template molecule (Compton 1991). Each new molecule is now available as template for the AMV reverse transcriptase in the so-called ´cyclic´ phase of NASBA. The RNA of the resulting RNA: DNA hybrid is hydrolysed by RNase H and extension of primer 1 and 2 again leads to the synthesis of a

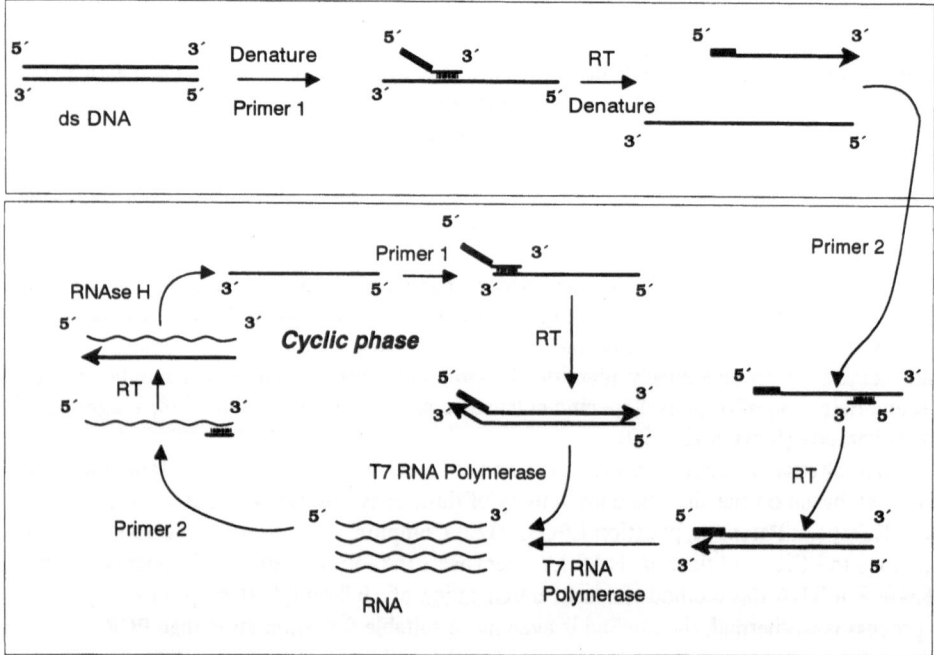

Figure 1. Principle of NASBA for amplification of DNA. The first box represents the PreNASBA reaction, where ds-DNA is transformed in ss-cDNA carrying a T7 promoter. The second box represents the NASBA reaction, where the ss-cDNA / T7 promoter is transformed in a ds cDNA with a functional promoter. Subsequently, RNA is transcribed by T7 RNA polymerase. Each new RNA molecule is now available as template for reverse transcriptase in the so-called ´cyclic´ phase of NASBA. RT = AMV reverse transcriptase. Boxes represent T7 promoter, solid lines represent DNA, wavy lines RNA and lines with arrrowed heads newly synthesized DNA.

functional promoter. NASBA is performed as a continuous, homogeneous and isothermal reaction in a single tube.

METHODS

Listeria monocytogenes DNA was isolated by a method of Flamm et al. (1984).Listeria cell lysates were prepared by heating of cell suspensions for 5 min in lysis buffer (final concentration: 1% (w/v) Triton X-100) at 90°C. The mixture was then centrifuged (5 min, 16.000 g) and 1μl of the supernatant was used for NASBA. The detection of *Listeria monocytogenes* in milk by immunomagnetic separation was essentially carried out as described by Skjerve (1990). Monoclonal antibodies produced against a cell bound antigen from *L.monocytogenes* were kindly provided by Arun K. Bhunia and Michael G. Johnson, Department of Food Science, Arkansas, USA (Siragusa and Johnson 1990; Bhunia et al 1991). The cells bound to the beads were washed two times with PBS (8 mM Na_2HPO_4, 2 mM KH_2PO_4 150 mM NaCl, 3 mM KCl, pH 7.4) containing 0,1 % RSA and 0,1% Tween, and resuspended in 50μl of PBS leading to a 20 x fold concentration. Listeria cell lysates were prepared as described. A typical PreNASBA reaction was performed in a 20 μl reaction mixture containing 40mM Tris, pH 8.3, 20mM $MgCl_2$ 40mM KCl, 5mM DTT, 1mM each dNTP, 2mM each NTP, 10U Sequenase, 0,2 M primer 1 and denatured target DNA. The reaction was incubated at 37°C (Sequenase) or 42 °C (AMV-RT) for 15 min, boiled and chilled on ice. One tenth of Pre-NASBA was added to the NASBA mixture.

NASBA was performed essentially as described by Kievits et al, (1991). A set of oligonucleotides was prepared to prime the amplification of a fragment of the PIII gene from *Listeria monocytogenes* (Boehringer Mannheim GmbH, International Patent Application, WO 90/08197) which is upstream of the lmaA gene (Göhmann et al 1990). The sequence of primer 1 was 5′- ATTCTAATACGA-CTCACTATAGGGAGACGCTTTACCTGCTTCGG CGATT-3' and of primer 2 was 5′-GTAATCATCCGAAACCGCTCA-3′ (expected fragment size: 181 nts).

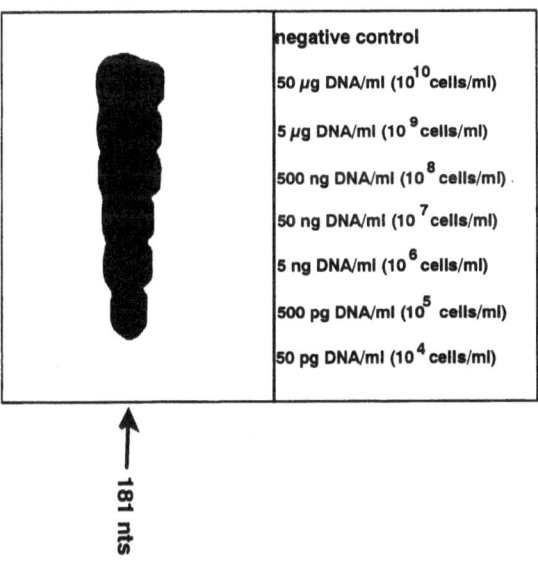

Figure 2. Sensitivity of the NASBA assay in detecting DNA from *L.monocytogenes* 1/2 a. NASBA products were analyzed by hybridization with an internal PIII-specific probe.

For detection of amplicons by Southern blot, 10 µl of NASBA mixture was run on a 1% agarose gel. The RNA was blotted onto a Nylon membrane using a vacuum-blot apparatus and hybridized with a digoxigenin-labeled oligonucleotide (5'-GTTTTACTTCTTGGACCG-3'). Hybrids were chemiluminometrically detected as described by Höltke et al (1992). For detection of amplicons by microtiter plate sandwich hybridization Dig-UTP (10µM final concentration) was incorporated into the nucleic acids by amplification. DIG-labeled amplicons were mixed with 10 ng of biotinylated oligonucleotide (same sequence as digoxygenin labeled oligonucleotide) in a total volume of 200 µl hybridization buffer (50 mM sodium phosphate buffer, pH 6.8, 750 mM sodium chloride, 0,05% BSA and 0,1% NP-40). The nucleic acid mixture was boiled for 5 min and hybridized at 37°C for 2 h in a streptavidin-coated microtiter plate. The hybrids were detected essentially as described by Eberle and Seibl (1992).

RESULTS

Sensitivity of NASBA

In order to determine the sensitivity of NASBA, dilutions of chromosomal DNA of *Listeria monocytogenes* were amplified as described in 'Methods'. As shown in figure 2 a NASBA product of the expected size (181 nts) was detected by Southern blotting with as little as 500 pg of *L. monocytogenes* DNA per ml (1×10^5 cells/ml). This corresponds to a detection limit of 100 analytes per amplification reaction. A similar experiment was performed with crude lysates of Listeria cells from 10^8 cells/ml down to 10^3 cells/ml. The sensitivity of this experiment was the same as for isolated DNA and could detect 10^5 cells/ml of Listeria monocytogenes.

Development of a microtiter plate sandwich hybridization assay

We have developed a microtiter plate sandwich hybridisation assay to identify specific NASBA products (Figure 3). The method does not include electrophoretic separation, is useful for the routine laboratory, employs standard laboratory equipment and reagents and should in principle allow automation.

Amplification and detection were carried out as described in 'Methods'. The assay was able to detect as few as 1×10^5 targets/ml which was equivalent to the Southern blot sensitivity. An additional advantage is the semiquantitative nature of the assay, so that the relative amount of original bacterial DNA can be monitored.

Detection of *L. monocytogenes* in milk

Listeria monocytogenes cells, diluted in milk, were captured and concentrated using magnetic beads coated with an antibody which reacts with a *L. monocytogenes* surface antigen. The immunocaptured cells were lysed and the DNA was subsequently used in NASBA.

The use of this technique provides a highly specific tool to capture the target bacteria from crude samples, with the additional advantage that the target can be concentrated and NASBA inhibitors are eliminated in a simple separation step. The detection limit was improved by this procedure to 5×10^3 cells/ml. This was in concordance with the fact that we reached a 20-fold concentration of cells by immunomagnetic separation.

Specificity of NASBA

To define the specificity of the described sandwich hybridization assay, NASBA was applied to different bacterial strains by using crude cell lysates of 10^7 CFU as template. The results obtained are presented in table 1. The method allows specific identification of *L.*

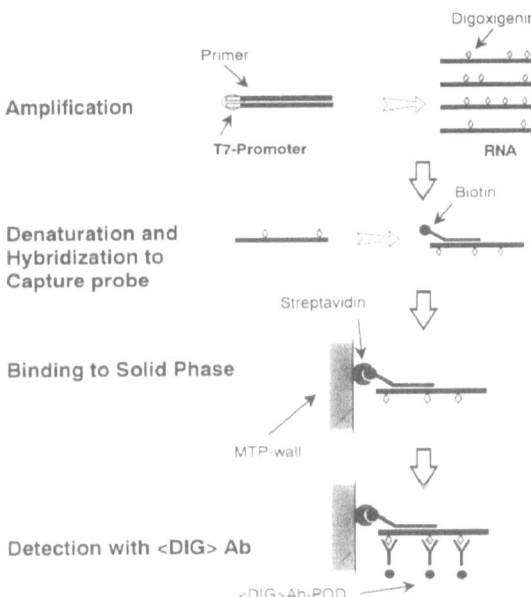

monocytogenes except for the serotypes 4ab and 3a, which are propably not recognized by the PIII primers. It is noteworthy that more than 90% of *L.monocytogenes* isolates from human and animal sources are represented by serotypes 1/2a, 1/2b and 4b (Balows et al., 1991). A positive reaction was not obtained from the 5 non-*L.monocytogenes* serovars and 6 strains of other genera.

CONCLUSIONS

- The PIII gene of the dth18 operon from *Listeria monocytogenes* allows species specific detection of this organism.
- The lower detection limit of NASBA for Listeria DNA is in the range of 10^5 targets/ml (hundred molecules of input).

- The combination of NASBA with an immunomagnetic capture of the target organism improves the detection limit to 5×10^3 targets/ml and minimizes sample influences on the amplification system.

- The microtiter plate sandwich hybridisation assay we developed offers a fast and reliable non-radioactive detection method for nucleic acid amplification products and is at least as sensitive as gel electrophoresis and blotting but is much more convenient and can be easily automated.

- The whole assay can be performed in four to six hours and allows rapid identification of the organism of interest.

Table 1. Specificity of *Listeria monocytogenes* sandwich hybridization assay. PreNASBA reaction with AMV-RT was performed at 42°C to increase specificity. Only one representative of each existing serotyped Listeria strain was tested.

Species	Serovar	Signal
Listeria monocytogenes	1/2 a	+
	1/2 b	+
	1/2 c	+
	3 a	-
	3 b	+
	3 c	+
	4 a	+
	4 b	+
	4 c	+
	4 d	+
	4 e	+
	4 ab	-
	7	+
L. ivanovii	5	-
L. innocua	6 a	-
	6 b	-
	4 ab	-
L. grayi		-
L. seeligeri		-
L. welshimeri		-
Erysipelothrix rhusiopathiae		-
Brochothrix thermosphacta		-
Kurthia zopfii		-
Enterococcus faecalis		-
Lactobacillus casei		-
Yersinia enterocolitica		-

ACKNOWLEDGEMENT

The authors appreciate the help of T. Chakraborty for the selection of DNA probes and of L. Malek for providing NASBA reagents and protocols. We thank Arun K. Bhunia and Michael G. Johnson for the supply of the antibodies.

REFERENCES

Balows A, Hausler Jr., WJ, Herrmann KL, Isenberg HD and Shadomy HJ. Listeria and Erysipelothrix. In Manual of clinical microbiology, 1991, p. 292. American Society for microbiology, Washington, D.C.

Bhunia AK, Ball PH, Fuat AT, Kurz BW, Emerson JW and Johnson MG. Development and Characterization of a Monoclonal Antibody Specific for *Listeria monocytogenes* and *Listeria innocua*. Infect Immun 1991, 59: 3176-3184

Compton J. Nucleic acid sequence-based amplification. Nature 1991, 350: 91 - 92

Eberle J. and Seibl R. A new method for measuring reverse transcriptase activity by ELISA. J Virol Methods 1992, 40 : 347 - 356

Flamm RK, Hinrichs DJ and Thomashow MF. Introduction of pAM1 into *Listeria monocytogenes* by conjugation and homology between native *L. monocytogenes* plasmids. Infect Immun 1984, 44: 157 - 161

Göhmann SM, Leimeister-Wächter E, Schiltz E, Goebel W. and Chakraborty T. Characterization of a *Listeria monocytogenes*-specific protein capable of inducing delayed hypersensitivity in Listeria-immune mice. Mol Microbiol 1990, 4: 1091 - 1099

Gray ML and Killinger AH. *Listeria monocytogenes* and listeric infections. Bacteriol Rev 1966, 30: 309 - 382

Höltke HJ, Sagner G, Kessler C and Schmitz G. Sensitive chemiluminescent detection of digoxigenin labeled nucleic acids: A fast and simple protocol and its applications. BioTechniques 1992, 12: 104 - 113

Innis MA, Gelfand DH, Sninsky JJ and White TJ. Diagnostics and forensics. In PCR protocols, a guide to methods and application, 1990 p.p. 399 - 406. Academic Press, San Diago

Kievits T, van Gemen B, van Strijp D, Schukking R, Dircks M. Adriaanse H, Malek L, Sooknanan R and Lens P. NASBA isothermal enzymatic in vitro nucleic acid amplification optimized for the diagnosis of HIV-1 infection. J Virol Methods 1991, 35 : 273 - 286

Schlech W. F., Lavigne PM, Bortolussi RA, Allen AC, Haldene EV, Wort. AJ, Hightower AW, Johnson SE, ing SH, Nicholls ES and Broome CV. Epidemic listeriosis-evidence for transmission by food. N Engl J Med 1983, 308: 203 - 206

Schwartz B. Hexter D, Broome CV, Hightower AW, Hirschhorn RB, Porter JD, Hayes PS, Bibb WF Lorber B and Faris DG. Investigation of an outbreak of listeriosis: new hypotheses for the etiology of epidemic *Listeria monocytogenes* infections. J Infect Dis 1989, 159: 680 - 685

Siragusa GR and Johnson MG. Monoclonal antibody specific for *Listeria monocytogenes*, *Listeria innocua*, and *Listeria welshimeri*. Appl Environ Microbiol 1990, 56: 1897 - 1904

Skjerve E, Rørvik LM and Olsvik Ø. Detection of *Listeria monocytogenes* in foods by immunomagnetic separation. Appl Environ Microbiol 1990, 56: 3478 - 3481

PCR and Virological Problems

CONFIRMATION OF HEPATITIS C VIRUS CARRIER STATE BY POLYMERASE CHAIN REACTION IN CASES WITH NONDIAGNOSTIC SEROLOGIES

Ch.-H. Wang and B. Flehmig

Department of Medical Virology and Epidemiology of Virus Diseases, Hygiene-Institute, University of Tübingen, Germany

Because of the heterogenous genome (quasi-species) and the glycosylated capsid proteins, hepatitis C viruses (HCV) may have a wide spectrum of variation in viral epitopes and therefore induce a very complex immune response. Although the specifity and sensitivity of the second generation of the anti-HCV ELISA have been improved, false-positive or false-negative results were found when the same test was applied in autoimmune diseases or rheumatoid arthritis or paraproteinemia or in HCV-infected individuals with delayed sero-conversion. Recently it was shown that an HCV infection can persist without detectable antibodies. We screened 400 anti-HCV-negative sera tested using the Abbott test kits as well as two other commercially available anti-HCV test kits. These sera had been submitted for HCV diagnosis from the university hospitals of Tübingen. 22 serum specimens (5.5%) with repeatedly nondiagnostic results in the different kits were found. The HCV-specific polymerase chain reaction (PCR) technique was used to search for the presence of HCV RNA in these specimens. Interestingly, 6 of 22 serum specimens were found to be HCV-PCR positive. Our results demonstrated that PCR remains the reference assay to clarify nondiagnostic serological results and to detect HCV infection in patients with no detectable anti-HCV immune response.

INTRODUCTION

HCV is the major etiologic agent of non-A, non-B hepatitis worldwide. It has a positive-stranded RNA genome of approximately 9500 nucleotides which encodes a polyprotein that is processed into structrual and nonstructrual proteins. HCV resembles flaviviruses and pestiviruses in genome organization within the polyprotein region and is proposed to be a member of the Flaviviridae (Alter et al 1989; Choo et al 1990, 1991; Han et al 1991; Houghton et al 1991).

Methods in DNA Amplification, Edited by
A. Rolfs *et al.*, Plenum Press, New York, 1994

HCV isolates show considerable amino acid sequence variation in coding regions, but they can be segregated into at least four different groups based on distinct amino acid sequence pattern (Houghton et al 1991; Okamoto et al 1992). The 5' untranslated region (UTR) of full-length HCV RNA appears to be 341 nucleotides long, at least based on putative full-length HCV clones reported to date (Chen et al 1992; Han et al 1991; Tanaka et al 1992; Okamoto et al 1991). Unlike the polyprotein region, the 5' UTR of HCV isolates are highly conserved (>98% within a group or >93 % between groups).

The development of a specific antibody assay with recombinant DNA technology has made diagnosis of hepatitis C virus (HCV) infection possible. Seroepidemiologic studies in various parts of the world have revealed that 60 to 100 % of patients with chronic non-A, non-B hepatitis have circulating anti-HCV antibodies. These studies identified HCV as the major etiologic agent of chronic posttransfusion non-A, non-B hepatitis worldwide. However, in acute non-A, non-B hepatitis, the frequency of detectable anti-HCV antibodies was much lower, ranging from 15 to 60%, with a delayed sero-conversion almost 3 to 6 months after infection. The delay in antibody response implies that some blood donors capable of transmitting HCV will not be detected by anti-HCV testing. A direct assay for HCV nucleic acids is thus useful in documenting HCV infection prior to anti-HCV seroconversion. Furthermore, when the same antibody test was applied to sera of the patients with autoimmune diseases and paraproteinemia, false-positive results were found (Teilmann, et al 1990; Ikeda et al 1990; Boudart et al 1990). As expected, Wang et al (1992) and Lazizi et al (1992) were able to detect infectious HCV genome in serum specimens without measurable anti-HCV-antibodies.

PCR has been shown to be a direct and more sensitive test than immunological tests detecting an infection and can be considered the most reliable criterion for distinquishing between true and false anti-HCV reactivity. In this study, a hepatitis C virus-specific polymerase chain reaction (HCV-PCR) was developed. We report here the result of HCV-PCR analysis of serum specimens with nondiagnostic serology.

MATERIALS AND METHODS

Serum samples and serological assays

400 human sera were drawn from patients in whom HCV serology was requested. These sera were tested by Abbott's anti-HCV testkit (second generation). All of the sera were shown to be anti-HCV negative. In order to confirm the negative results, the sera were further tested by two other test kits (Wellcome & GBC). The three different anti-HCV kits contained different viral antigen components for anti-HCV antibody capture. All the serological assays used and the calculation of the results were performed in accordance with the intructions of the manufacturer. The cut-off value was calculated by adding 0,6 to the mean of the extinction [E450] of the negative control replicates. A test result where the E450 is equal to or greater than the cut-off value indicates of the presence of antibody. Such specimen were retested in duplicate using the original sample source. Specimens giving a repeatedly positive result were presumed to contain antibody to HCV antigens and were further investigated. (Cut-Off Index = measured value/Cut-Off value; Cut Off Index: > 1, sample is positive; < 1 sample is negative).

RNA preparation

All the samples were stored at -20°C until the PCR test was performed. RNA from serum or plasma was extracted as followed. Briefly, 100 μl of serum or plasma was incubated with 200 μg of proteinase K per μl containing 0,2 M Tris-HCl (pH 7.5), 25 mM EDTA, 0,3 M NaCl, 2% sodium dodecyl sulfate at 37°C for 40 min. After phenol-chloroform extraction and ethanol precipitation, the pellet was dissolved in water. RNA from 20 μl of serum was reverse transcribed in a 20 μl mixture containing the outer antisense primer (0,6 μM), 50 mM dithiothreitol, 0,5 mM (each) deoxynucleoside triphosphate (dNTP), 40 U of RNAsin (40,000 units/ml, Promega, USA), and 15 U of AMV reverse transcriptase (10,000 units/ml, Promega, USA) at 43°C for 45 min.

HCV-specific polymerase chain reaction

PCR was performed with nested primers based on the highly conserved 5' non-coding region (Table 1). The sequence of the outer primers (TÜ1-TÜ2) and the inner primers (TÜ3-TÜ4) was synthesized by TIB INC. (Berlin, FRG). cDNA was heat-denatured at 95°C for 5 min and chilled in ice. Amplification of HCV cDNA was performed by the following method. Briefly, 100 μl of the reaction micture containing 5 μl of specimen cDNA, 50 mM Tris-HCl (pH (8,0), 2,5 mM MgCl$_2$, 500 pM (each) outer primers (TÜ1-TÜ2) and 2U of Tag DNA polymerase (Perkin Elmer-Cetus) was overlaid with 100μl of mineral oil. 40 cycles were done at 94°C for 15s, 54°C for 15s, 74°C for 15s. After the first round of amplification the samples were submitted to a second round of amplification with the inner primers (TÜ3-TÜ4) under the conditions described above.

Analysis of PCR amplification products

An aliquot of 1/10th of the PCR amplification products was analysed on a 2% agarose gel, stained with ethidium bromide, or blotted onto nitrocellulose membrane, and hybridized to a digoxigenin-labelled HCV probe.

Hybridization of PCR products

The specificity of the amplified products were confirmed by hybridization with a digoxigenin (DIG)-labeled oligonucleotide. Principally, the oligonucleitide probes are detected, after hybridization to target nucleic acids by enzyme-linked immunoassay using antibody-conjugate (anti-DIG perioxidase conjugated, anti-DIG-POD). A subsequent enzyme-catalyzed color reaction visualizes hybrid molecules.

At first, HCV-specific oligonucleotide was labelled with DIG using a DIG-oligonucleotide 3'- end labelling kit (cat. Nr. 1362372, Boehringer Mannheim, Germany). 100 μl of PCR products were slot-blotted on the nitrocellulose membrane (NM) by baking. The NM was pre-hybridized in a sealed plastic bag or box with at least 20 ml hybridization solution per 100 cm^2 NM at 42°C for at least 1 hr. The solution was replaced with about 2,5 ml of hybridization solution (containing 2μl of freshly DIG-labelles oligonucleotide) per 100cm of NM. The NM was incubated for at least 6 hours at 42°C with redistribution the solution occasionally. The NM was washed 2 x 5 min at room temperature with at least 50 ml of 2 x SSC buffer (NaCl,

300 mM, Na-citrate, 30 mM; pH 7.0, 25ºC), SDS, 0,1% (w/v), per 100 cm² NM and 2 x 15 min at 68ºC with 0.1 x SSC, 0.1% (w/v). Thereafter the NM was stored air-dried for later use. For the enzymatic color detection of DIG-oligonucleitide hybridized NM, an anti-DIG-POD diluted in 20 ml PBS (1:10,000) was incubated with the NM for 30 min. Unbound conjugate was removed by washing 2 x 15 min with 100 ml washing buffer. The reaction on the NM was visualized by addition of substrate for peroxidase.

Table 1. Oligonucleotide sequences of primers used for amplification of HCV RNA. Primer TÜ1 was used for cDNA synthesis and PCR. The product size of primers TÜ1 and TÜ2 was 271 bp; that of TÜ3 and TÜ4 was 256 bp, nucleotide positions are numered.

Primer	Polarity	Sequences (5'to3')	Nucleotide Position
(A):Outer Primers:			
TÜ1	sense		-289/-264
		5'-AACTA'CTGTC'TTCAC'GCAGA'AAGCG'-3'	
TÜ2	antisense		- 18/-43
		5'-CTCCC'GGGGC'ACTCG'CAAGC'ACCCT'-3'	
(B):Inner Primers:			
TÜ3	sense		-279/-256
		5'-TTCAC'GCAGA'AAGCG'TCTAG'CCATG'-3'	
TÜ4	antisense		-25/-51
		5'-ACTCG'CAAGC'ACCCT'ATCAG'GCAGT'-3'	

RESULTS

Controversial results

400 serum samples which were repeatedly negative by Abbott anti-HCV kits (TK1; second generation) were selected for our study. These sera had been submitted for HCV diagnosis from the university hospitals of Tübingen. In order to confirm the negative serologic results we examined these serum samples with two other anti-HCV kits (TK2 and TK3). As can be seen in Table 2, 22 sera were found to be anti-HCV positive. We repeated this assay two times and the same results were obtained. The samples with conflicting results in three assays were selected for further analysis by the anti-HCV immunoblotting assay by RIBA II (Chiron). All of these specimens with conflicting serological results were judged to be anti-HCV indeterminate by RIBA II or negative. The contradictory results between TK1 (the second generation assay of Abbott), TK2 (Wellcome), TK3 (GBC) and RIBA were therefore investigated by hepatitis C virus-specific polymerase chain reation (HCV-PCR).

Table 2. PCR analysis of 22 specimens with conflicting serological results tested by three serological assays and RIBA-: negative results; +: positive results; IND: Indeterminate result cut off index was determined as described in MATERIALS & METHODS (=measured value/cut-off value). TK1, TK2 and TK3 are testtest kits obtained from Abbott (USA), Wellcome (England) and GBC (Taiwan, ROC), respectively. [1, 2, 3] represent measured extinction value (E450).

Serum#	TK1	TK2	TK3	RIBA	PCR
1	0,181[1]	0,981[2]	1,652[3]	IND	-
2	0,272	1,382	0,297	-	-
3	0,153	1,269	1,394	IND	+
4	0,247	1,071	0,307	-	-
5	0,228	1,004	0,015	IND	+
6	0,247	5,934	0,035	IND	-
7	0,194	0,486	0,019	IND	+
8	0,203	1,259	0,204	-	-
9	0,272	1,708	0,488	-	-
10	0,040	1,161	0,267	-	-
11	0,423	1,014	0,503	-	-
12	0,194	1,051	0,457	-	-
13	0,495	1,372	0,247	-	-
14	0,307	1,202	0,194	IND	-
15	0,219	1,122	0,347	-	-
16	0,357	1,009	1,040	IND	+
17	0,219	1,363	0,156	-	-
18	0,203	0,745	1,216	-	-
19	0,338	1,099	0,217	IND	-
20	0,755	1,618	2,429	IND	+
21	0,498	4,632	2,204	IND	+
22	0,529	1,278	0,890	-	-

The specificity of HCV-PCR

The products of the first round of amplification, which corresponded to a size of 271 bp, were only weakly or not at all detectable in most samples, in contrast to the positive control. After the second round, the amplified PCR samples which corresponded to an expected size of 256 bp were clearly visible in 6 of the 22 serum samples. In Figure 1, five of them are shown in the agarose gel. As control for our PCR analysis, five HCV-positive and five HCV-negative sera were used. With the same experimental conditions, HCV RNA was detected by PCR in all those positive serum samples that were anti-HCV positive (data not shown). Additional specificity controls, such as RNA samples extracted from sera from several healthy humans and from humans with hepatitis B virus infections, were included in order to avoid false-positive results because of contamination. A control with no DNA reagent (i.e. PCR mixture without template DNA) was also used during all the steps of the protocol.

The specificities of the amplified products were confirmed by hybridization with a digoxigenin (DIG)-labelled oligonucleotide probe corresponding to positions -91/-51 of the 5' non-coding region. For this assay, PCR products were first slot-blotted and hybridized with the DIG-labelled HCV specific oligonucleotides. As can be seen in Figure 2, twenty-four PCR positive products (lines 1, 2, 3) and twenty-four negative products (line 4, 5, 6) from different sera were shown. All of the HCV-negative samples were negative as well as the negative control.

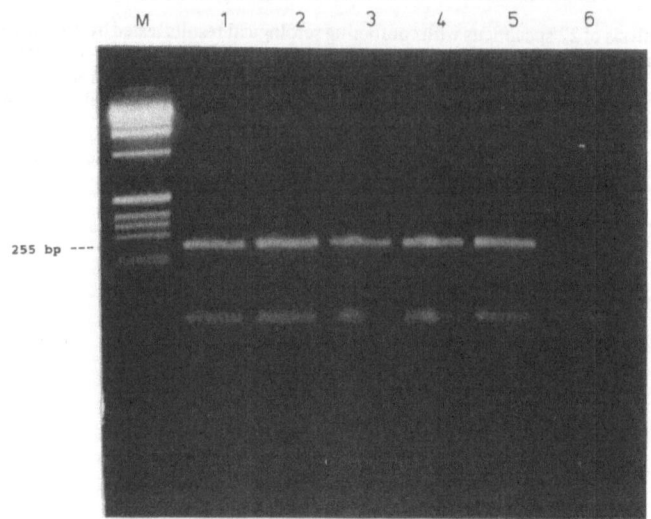

Figure 1. Detection of hepatitis C virus with nested polymerase chain reaction: Ethidium bromide-stained gel showing the PCR product bands obtained after second round of amplification. MW: molecular size marker, Lane 1, 2, 3, 4, 5:HCV-PCR of five infected sera in different HCV antigen carriers. Lane 6: Negative control sera

In contrast, the RNA genome of all of the HCV positive sera was amplified by the PCR assay with our primers (TÜ1-TÜ4). The results demonstrate that the described PCR assay is very specific and suitable for HCV-RNA detection. The assay was employed to analyse 22 specimens with conflicting serological results. As a result, under identical experimental conditions, 6 of 22 specimens were demonstrated to be HCV-PCR positive.

DISCUSSION

Because the antigen components for the detection of the anti-HCV in the three commercial test kits are different, it is possible that these different antigen components can react with different types of antibodies. The controversial results which are found in our study can therefore on the one hand be explained by these different antibody reactions. On the other hand, the antibody response against different molecules of HCV may vary from individual to individual as in the case of hepatitis A viral infection (Wang et al 1993) and may be influenced by pathological conditions of the host's immune system. It is also conceivable that these patients were still in the window of seroconversion. Clinical significance of the presence of HCV genomes - in particular their relevance for the host's immune response - should be clarified in the near future.

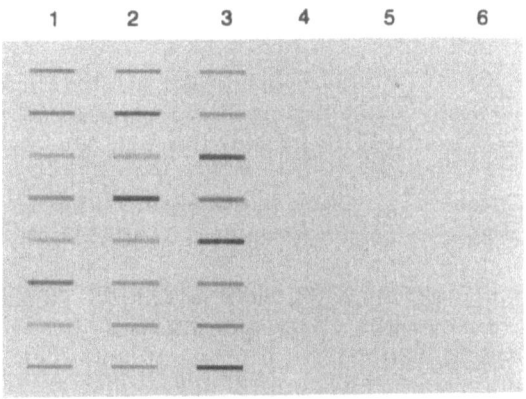

```
1,2,3:  HCV-PCR  positive  sera
4,5,6:  HCV-PCR  negative  sera
```

Figure 2. Specificity of the HCV-PCR detection with primers (TÜ1-TÜ4). PCR products of the HCV specific genome were slot-blotted on the nitrocellulose membrane and hybridized by a digoxigenin-labelled HCV-specific oligonucleotide probe. Line 1,2,3: PCR products of twenty-four HCV-PCR positive sera derived from the 6 infected patients. Line 4,5,6: PCR products of twenty-four HCV-PCR negative sera derived from the other 16 patients.

Because of the variation of the RNA genome and the glycosylated capsid proteins, HCV is thought to induce a wide spectrum of different immune responses (Martell, et al., 1992; Weiner, et al., 1992). In many cases of the recombinant-produced viral antigens, the antigenicity induced in the hosts were restricted to cloned conformation (Lemon et al., 1992). Moreover, viral epitopes other than the recombinant viral epitopes or the synthetic peptides of the test kits (5-1-1, c100-3, c-33c, c22-c, NS4 and NS5 regions) may exist. Therefore, it may be also possible that the antigens used in the test kits of our serological laboratory did not mimic the authethic viral epitopes which produced different antibodies after immune response and may explain why the test kits failed to detect antibodies. Further identification of the sequences of the strains detected could eventually explain why the strains do not induce a detectable immune response.

Currently, antibody testing for HCV establishes whether an individual has responded immunologicaly to exposure to viral antigens but does not determine whether the infectious agent has been eliminated or continues to persist in the host. Direct detection of HCV genomes is a better marker and useful to distinquish between infectious individuals and immune individuals. With this method, one can identify viral nucleic acid in the clinical course of infection and the significance of HCV antibodies. In summary, PCR has to be considered as a reference assay to discriminate serology results due to low level of antibodies as well as a reference assay to be used when serology is nondiagnostic or when false positivity is suspected.

ACKNOWLEDGEMENT

We indebted to Prof. Dr. H.-G. Gerth for support and helpful discussion of the manuscript. We also thank Dr. S.-Y. Tschen for designing and typing the manuscript.

REFERENCES

Alter, HJ, Purcell, RH, Shih, JW, Melpolder, JC, Houghton, M, Choo, QL and Kuo, G. Detection of antibody to hepatitis C virus in prospectively followed transfusion recipients with acute and chronic non-A, non-B hepatitis. N. Engl. J. Med. 1989, 321:1491-1500

Boudart, D, Lucas, JC, Muller, JY, Le Carrer, D, Planchon, B, Harousseau, JL. False-positive hepatitis C virus antibody tests in paraproteinaemia, Lancet 1990, 336:63-64

Chen, PJ, Lin, M., Tai, KF, Liu, PC, Lin, PC, Lin CJ and Chen DS. The taiwanese hepatitis C virus genome: Sequence determination and mapping the 5'terminus of viral genome and anti-genomic RNA. Virology 1992, 88:102-113

Choo, QL, Richman, KH, Han, JH, Berger, K, Lee, C, Dong, C, Gallegos, C, Coit, D, Medina-Selby, A, Barr, PL, Weiner, AJ, Bradley, DW, Kuo, G, Houghton, M. Genetic organization and diversity of the hepatitis C virus. Natl. Acad. Sci. USA 1991, 88:2451-2455

Han, JH, Shyamala, V, Richman, KH, Brauer, ML, Irvine, B., Urdea, MS, Tekamp-Olson, P, Kuo, G, Choo, QL and Houghton, M. Characterization of the terminal regions of hepatitis C viral RNA:Identification of the conserved sequences in the 5'untranslated region and poly(A) tails at the 3'end. Proc. Natl. Acad. Sci. USA 1991, 88:1711-1715

Houghton, M, Weiner, A, Han, J, Kuo, G and Choo, QL. Molecular biology of the hepatitis C viruses:Implication for diagnosis, development and control of viral disease. Hepatology 1991, 14:381-388

Ikeda, Y., Toda, G., Hashimoto, N., Kurokawa, K. Antibody to superoxide dismutase, autoimmune hepatitis, and antibody tests for hepatitis C virus. 1990, 335:1345-1346

Lemon, SM, Barclay, W, Ferguson, M, Murphy, P, Jing, L, Burke, K., Wood, D, Katrak, K, Sangar, D, Minor, PD and Almond JW. Immunogenicity and antigenicity of chimeric Picornaviruses which express hepatitis A virus (HAV) peptide sequences:Evidence for neutralization domain near the amino terminus of VP1 of HAV. Virology 1992, 188:285-295

Lazizi, Y, Elfassi, E and Pillot, L. Detection of hepatitis C virus sequences in sera with controversial serology by nested polymerase chain reaction, J. Clin. Microbiology 1992, 30:931-934

Martell, M, Esteban, JI, Quer, J, Genesca, J, Weiner, A, Esteban, R, Quardia, J and Gomez, J. Hepatitis C virus (HCV) circulates as a population of different but closely related genomes:quasispecies nature of HCV genome distribution. J.Virology 1992, 66:3225-3229

Okamoto, H, Kurai, K, Okada, SI, Yamamoto, K, Lizuka, H, Tanaka, T, Fukuda, S, Tsuda, F. and Mishiro, S. Full-length sequence of a hepatitis C virus genome having poor homology to reported isolated : comparative study of four distinct genotypes. Virology 1992, 188:331-341

Tanaka, T, Kato, N, Nakagawa, M, Ootsuyama, Y, Cho, MJ, Nakazawa, T, Hijikata, M, Ishimura, Y. and Shimotohno, K. Molecular cloning of hepatitis C virus genome from a single Japanese carrier: Sequence variation within the same individual and among infected individuals. Virus Research 1992, 23:39-53

Theilmann, L, Blazek, M, Goeser, T, Gmelin, K, Kommerell B. and Fiehn, W. False-positive anti-HCV tests in rheumatoid arthritis Lancet 1990, 335:1346

Wang, CH, Fisher, B. and Flehmig, B. Diagnosis of hepatitis C virus in blood samples with ELISA, RIBA, and PCR, NEWS LETTER of European Group for Rapid Viral Diagnosis 1992, 23:25

Wang, CH, Heinricy, U, Weber, M, Pfisterer, M, and Flehmig, B. Acute and long-term immune response against viral capsid proteins after natural hepatitis A virus (HAV) infection. in "Viruses and virus-like agents in disease", a Karger Symposium in Basel, Switzerland, Karger Publishers, 1993, p35

Weiner, AJ. Geysen, HM, Cristopherson, C, Hall, JE, Marson, TJ, Saracco, G, Bonino, F, Crawford, K, Marion, CD, Crawford, KA, Brunetto, M., Barr, PJ, Miyamura, T, McHutchinson, J, and Houghton, M. Evidence for immune selection of hepatitis C virus (HCV) putative envelope glycoprotin variants: Potential role in chronic HCV infections. Proc. Natl. Acad. Sci. USA 1992, 89:3468-3472

Yoo, BJ, Spaete RR, Geballe AP, Selby M, Houghton, M. and Hahn, JH 5'End-dependent translation initiation of hepatitis C viral RNA and the presence of putative positive and negative translational control elements within the 5'untranslated region. Virology 1992, 191:889-899

CONFIRMATION OF HEPATITIS C VIRUS POSITIVE SERAS BY POLYMERASE CHAIN REACTION

G. Lucotte[1], C. Mura[2], A. Aouizenate[3], T. Champenois[3], and J. Marchand[4]

[1]Laboratory of Molecular Anthropology, CHU of Choin Port-Royal, [2]Laboratory INSERM Y 120, Robert Debré public Hospital, [3]Laboratory LCL, 37, Paris, [4]CIS BIO International, BP 32, Gif-sur-Yvette, France

INTRODUCTION

Following the initial discovery (Choo et al 1989) of a cDNA clone reactive in an elevated proportion of chronic non-A, non-B hepatitis (NANBH) patients, the genome of a RNA virus was identified as hepatitis C virus (HCV). Several serological tests are now available, notably two second-generation AIAs (Ortho Diagnosis and Abott Diagnostics), and two types of supplementary antibody assays are also available: a neutralisation assay (Abbott) based on the C100-3 antigen, and a recombinant immunoblot assay (Chiron-Ortho) with HCV structural (c22-3) an non-structural (c33; c100-3) antigens and the fusion protein (SOD) control.

Polymerase chain reaction (PCR) is extremely sensitive and specific in confirming the presence of the HCV genome (Simmonds et al 1990) when using primers in the highly conserved 5′ non-coding region (Okamoto et al 1990; Garson et al 1990). This high sensitivity prompted us to use PCR in serologically positive sera in an attempt to identify viral carriers. We demonstrate that PCR is an efficient method to confirm or establish viral charge in seras, on EIA positive and negative samples. Consequently, we propose that PCR could be used as the second test method, conjointly to EIA.

PATIENTS AND METHODS

All sera used in this study were obtained from volunteers coming to LCL. Most of them have abnormal alanine aminotransferase (ALT) values or hepatitis symptoms. On arrival all samples were stored at -20°C until serological testing and PCR (realized independently) were performed.

Methods in DNA Amplification, Edited by
A. Rolfs *et al.*, Plenum Press, New York, 1994

The second-generation EIA of Abott, containing the c100-3 antigen supplemented with recombinant HCV structural (c22-3) and non-structural (c33 from NS3) antigens were used at first. Results are expressed as S/Co ratios.

Samples were simultaneously prepared also for PCR: 100µl serum was mixed with 500µl of extraction solution [4M guanidium thiocyanate, 25 mM sodium citrate (pH = 7,0), 0.5 % sarcosyl, 0.1M ß-mercaptoethanol] an 10µg of E. coli derived carrier tRNA was then added per tube. Thereafter 100µl of 2M sodium acetate (pH = 4,0), 500µl of water saturated phenol, and 100 µl of chloroform-isoamylalcohol (49:1) was sequentially added per tube. Each tube was shaken for 10 sec, cooled on ice for 15 min., and the aqueous and phenol-chloroform phases were separated by centrifugation for 20 min. at 10,000 g and 4°C. The upper aqueous phase of each tube was transferred to fresh tubes and precipitated with 500 µl of isopropanol at -20°C for one hour, followed by centrifugation for 20 min. at 10,000 g and 4°C. The pellet was then dissolved at 0°C in 30µl diethylpyrocarbonate treated H_2O with 60 units of RNAse inhibitor, divided in 3 aliquots, and frozen at -70°C.

One aliquot for each sample was thawed on ice immediately before being added to 40 µl of fresh RNAse-free reaction mix. The final concentrations of the buffer were: Tris-HCl, 20 mM (pH = 8.4); $MgCl_2$, 2.5 mM; KCl, 50 mM; dNTP (Boehringer Mannheim), 0.25 mM each; RNAsin, 25 U/reaction; AMV reverse transcriptase (Promega), 5 U/reaction; Taq polymerase (Amplitaq, Cetus Perkin-Elmer), 0.5 U/reaction. Primers were all from the conserved 5′-moncoding region, and have been used elsewhere (Widell et al., 1991). The outer primers of the first PCR were 5′-CATGGTGCACGGTCTACGAGACC-3′(antisense) and 5′-GGCGACACTCCACCATAGATC-3′ (sense), both with final concentrations of 0.5 mM. After mixing RNA with reagents, mineral oil was used to prevent evaporation, and reverse transcription (RT) was performed at 43°C for 20 min. (Hybaid Thermal Reactor) and then 35 cycles of thermocycling with the same tube were started. Melting was done at 95°C for 1 min., annealing at 45°C for 2 min., and elongation at 72°C for 3 min.

The second, inner PCR, was performed with 5 µl form the first reaction transfered to 45 µl of the same buffer as above, excluding RT reagents and outer primers. The inner primers (0.5µM also) were 5′-TCGCAAGCACCCTATCAGGCAG-3′ (antisense) and 5′GGAACTACTGTCTTCACGCAGA-3′(sense). Twentyfive cycles of 95°C for 1 min and at 60°C for 1 min were run.

First and second round products were analysed by electrophoresis on a 1.5 % agarose, stained by ethidium bromide and visualised by UV transillumination. A fragment of 327 bp for the first round, and of 260 bp for the second round (Figure 1) was expected.

Every run included a weak diluted positive control plasma, a PCR reaction control, and a negative control plasma. All precautions were taken to avoid contamination (Kwoks and Higuchi 1989). Samples were processed coded and the sera, only positive in the second round, were repeated.

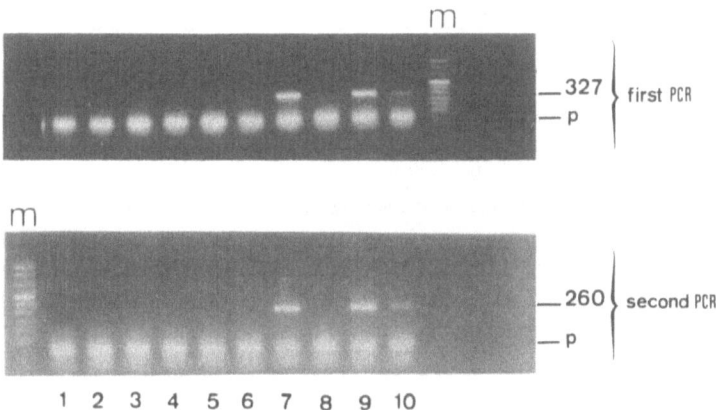

Figure 1. Above, first PCR, below, second PCR. m: fragment DNA molecular weight markers; p: residual primers observed in each of the 10 lames; expected bands are obtained at 327 bp for the first PCR, and at 260 bp for the second PCR (lanes 7,9 and 10 correspond to PCR > 0 samples).

RESULTS AND DISCUSSION

Table 1 summarizes the results that we obtained on about 2000 independent samples tested simultaneously for EIA and PCR. A number of 1636 of them are negative both for EIA and PCR; but 38 on a total of 2094 samples tested (1.8 %) are EIA negative but positive in PCR (often revealed during the second PCR). In these cases PCR appears a more sensible technique than the second-generation EIA used. On the other hand, on 420 samplex EIA positive, only 297 of them are PCR positive, about 1/3 of the total number of EIA positive samples (123 of them: 29.2 %) being PCR negative. All these discordant samples were successively tested twice for PCR negativity. We admit that these discordant samples correspond to seras with antibody persistence after viral clearing.

Table 1. Results obtained (N: numbers) with (%) on a total of 2094 samples tested simultaneously for: E: EIA an P: PCR.

	N (%)	E AND P
	1636	E- P-
	297	E+ P+
	123 (29.2)	E+ P-
	38 (1.8)	E- P+
Total	2094	

REFERENCES

Choo QL, Kuo G, Weiner AJ, Overby LR, Bradley DW, Houghton M. Isolation of a cDNA clone derived from blood borne non-A, non-B viral hepatitis genome. Science 1989, 244: 359-362

Garson JA, Ring C, Tuke P, Tedder RS. Enhanced detection by PCR of hepatitis C virus RNA. Lancet 1990, 336: 878-879

Kwok S and Higuchi R. Avoiding false positives with PCR. Nature 1989, 339: 237-238

Okamoto H, Okada S, Sugiyama Y, Tanaka T, Sugai Y, Akahane Y, Machida A, Mishiro S, Yoshizawa H, Miyakawa Y. Detection of hepatitis C virus RNA by a two-stage polymerase chain reaction with two pairs of primers deduced from 5'-non-coding region. Jap J Exp Med 1990, 60: 215-222

Simmonds P, Zhang LQ, Watson H, Rebus S, Fergunson ED, Balfe P, Leadbetter GH, Yap PL, Peutherer JF, Ludlam CA. Hepatitis C quantification and sequencing in blood products, haemophiliacs and drug users. Lancet 1990, 336: 1469-1472

Widell A, Månsson AS, Sundström G, Hansson BG, Nordenfelt E. Hepatitis C virus RNA in blood donor sera detected by the polymerase chain reaction. J Med Virol 1991, 35: 253-258

SPECIFIC DETECTION AND RAPID IDENTIFICATION OF HUMAN ENTEROVIRUSES BY PCR AMPLIFICATION

U.Kämmerer[1], B. Kunkel[1] and K. Korn[2]

[1]Medical Clinics II and [2]Institute of Clinical and Molecular Virology, University of Erlangen-Nürnberg, Germany

A nested PCR was developed that allows the specific amplification of enteroviruses as well as rhinoviruses with one set of four primers from the 5′-noncoding region. For rapid identification of the respective serotypes, restriction enzyme (RE) digestion of PCR products was done. Experiments with serotyped isolates showed a sensitivity of 10 genome equivalents or less. RE digestion could predict the serotype correctly in almost all cases investigated, minimizing the need for sequencing of PCR products. Application to clinical samples is demonstrated for cases of aseptic meningitis and other neurological diseases and for myocardial biopsies from experimentally infected animals as well as from patients with dilative cardiomyopathy.

INTRODUCTION

Picornaviruses are very important human pathogens. These small RNA viruses cause a wide variety of diseases, ranging from common cold over infections of the lower respiratory and intestinal tract to paralytic poliomyelitis. The more than 170 distinct serotypes isolated from humans include three polioviruses (PV 1-3), six coxsackie B viruses (CBV 1-6), 23 coxsackie A viruses (CAV 1-22,24), 31 echoviruses (EV 1-9, 11-27, 29-33), more than 100 human rhinoviruses (HRV), four serotypes of undesignated human enteroviruses (EN 68-71) and the hepatitis A virus (Stanway 1990).

The picornaviral genome is a single stranded, positive-sense RNA 7,200 - 8,500 nucleotides in length. At the 5′-end of the genome there is a region of 600 - 1,000 bases that is not translated (Rückert 1990). This so-called 5′-non-coding region contains several blocks of nucleotides which are highly conserved among the picornaviruses. These conserved sequences allow one to select short primers for a group-specific PCR making the detection of a wide variety of picornaviruses possible.

For the nested PCR (nPCR) shown here, four primers of 17 bp each in length are combined for an outer (first PCR) and inner (nPCR) primer set.

METHODS

RNA extraction: RNA from liquid specimens (virus strains, tissue culture supernants, CSF, sera, stool suspensions) was extracted following a protocol from Rotbart (1990). A 100µl sample was first mixed with 5.2µl of 10% sodium dodecyl sulfat (SDS) to a final concentration of 0.5%. After the addition of 105µl of phenol:chloroform:isoamyl alcohol (50:49:1), the mixture was vigorously vortexed and centrifuged at 13000 rpm for 5 min at 4°C. The aqueous phase was removed to a new tube and the remaining organic phase re-extracted with one volume solution A (10mM Tris/HCl, pH 7.5, 100mM NaCl, 1mM EDTA and 0.5% SDS). This mixture was again centrifuged under the above conditions and the resulting aqueous phase added to the first. In these aqueous phases, RNA was precipitated by addition of 1/3 vol. 8M ammonium acetate (final conc. 2M) and 3 vol. 100% ethanol for 2h at -20°C. RNA was sedimented by centrifugation at 13000 rpm at 4°C for 30 min. Supernant was discarded, the pellet washed once with 75% ethanol and dried under vacuum.

RNA extraction from tissues (heart biopsies, cell pellets) followed the acid guanidinium-thiocyanate-phenol-chloroform extraction by Chomcszynski and Sacchi (1987). Samples of 0.1-0.4 mg were extracted with 300µl of fresh solution D (4M guanidinium-thiocyanate, 25 mM sodium citrate, pH 7.0, 0.5% sarcosyl, 0.1 M 2-mercaptoethanol) followed by sequentially adding (and mixing with) 30µl 2M sodium acetate pH 4.0, 300µl phenol (water saturated) and 60µl chloroform-isoamyl alkohol (49:1). The resulting white mixture was cooled on ice for 15 min and then centrifuged at 10000xg at 4°C for 30 min. The aqueous phase was transferred to a fresh tube and the RNA precipitated with one volume isopropanol at -20°C for 2h. RNA was sedimented by centrifugation, washed and dried as described above.

Reverse transcription (RT) reaction: For RT reaction, the lyophilized RNA was resuspended in a 30µl reaction mixture containing 75 mM KCl, 50 mM Tris/HCl (pH 8.3), 3 mM $MgCl_2$, 10 mM dithiothreitol, 0.3 mM (each) deoxynucleotide triphosphates (Boehringer Mannheim), 90 ng of primer Coxprim 2 and 100 U of Moloney murine leukemia virus reverse transcriptase (MMLV-RT, Gibco-BRL). The mixture was incubated at 37°C for 60 min, followed by 5 min at 95°C to inactivate reverse transcriptase.

Nested PCR: For enzymatic amplification, 5µl of the RT-mixture was added to 95µl of PCR mixture containing: 50 mM KCl, 10 mM Tris/HCl (pH 8.3), 0.001% (w/v) gelatin, 1,5mM $MgCl_2$, 0.25 mM (each) deoxynucleotide triphosphates, 90 ng (each) of Coxprim 1 and 2 and 1.5 U of Taq DNA Polymerase (Promega). Fourty cycles of denaturation (94°C, 20 sec), annealing (45°C, 30 sec) and primer extension (72°C, 45 sec) were then performed in a thermal cycler. For nPCR, 2µl of the first PCR mixture were added to 98µl of the second PCR mixture (same as first, but Coxprim 3 and 4 instead of 1 and 2). After another set of 40 cycles under the above conditions, a 10µl aliquot was analyzed by agarose gel electrophoresis.

To avoid false-positive PCR results, the universal precautions for PCR described by Kwok and Higuchi (1989) were strictly followed.

Agarose gel electrophoresis: Ten microliters of PCR products were run on horizontal submerged, 1.5 % agarose gels in 0.5 x TBE (0.045 M Tris-borate, 0.001 M EDTA, pH 8.0) at 140 V for 1 h and stained with ethidium bromide. The DNA fragments resulting from RE-digestion were separated in 3% agarose gels under the above conditions.

Restriction enzyme (RE) digestion: For RE-digestion, amplified DNA was ethanol precipitated with 0.1 volume 3 M sodium acetate and 2.5 volume 96% ethanol at -20°C for 2h. After centrifugation at 13,000rpm at 4°C for 30 min, the pellet was washed with 75% ethanol, dried and resuspended in 50% of the original volume. A 5µl aliquot was incubated with 1-5 U of RE in a standard reaction with the proposed RE-buffer at 37°C for 1 h.

RESULTS AND DISCUSSION

A computerized database search showed that with exception of the hepatitis A virus, which is highly divergent in sequence, the primers Coxprim 1-4 (Table 1) should detect all human picornaviruses whose sequences are available in the Genbank and EMBL databases. No other known human or viral sequences have been detected in these databases by all four primers. The combination of Coxprim 1 and 2 (the latter of which was used as the antisense primer for reverse transcription of viral RNA to cDNA) results in an amplified fragment of about 493 bp in length. Coxprim 3 and 4 as a nested primer pair result in a PCR fragment of approximately 297 bp (Figure 1).

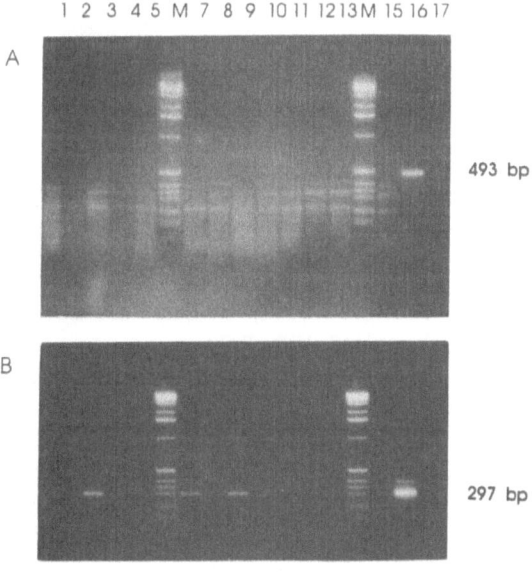

Figure 1. Comparison of intial (A: Coxprim 1+2) and nested (B: Coxprim 3+4) PCR from heart biopsies. Lanes 1-5, 7-13, 15: biopsy samples, lane 16: positive control (PVIII), lane 17: neg. control, M: 1 kb ladder

To confirm the assumption that the described nPCR allows the detection of most of (or maybe all) human picornaviruses with the exception of the hepatitis A virus, a wide range of picornaviral prototype strains (PV I (Mahoney), PV. II, PV III (Saukett), PV III (Leon 12a1b), CAV 5, 7, 9, 16, 21, CBV 1-6, EN 71, EV 1-9, 11-22, 24-27, 29-33 and HRV 1b, 2, 14, 89) and clinical isolates (PV I-III, CAV 9, 21, CBV 1-6 and EN 71) have been tested and all gave successful amplification with the nPCR. Because of these results and the fact that nPCR with Coxprim 1-4 gave no detectable amplification with uninfected Vero and HeLa cells, as well as with CMV, HSV, EBV and rotavirus, this nPCR is shown to be group-pecific for human enteroviruses and rhinoviruses.

Table 1. Comparison of Coxprim 1-4 with the sequences of that human enteroviruses and r h i n o v i r u s e s, available in the Genbank and EMBL databases.(PV= poliovirus, CAV= coxsackie A virus, CBV= coxsackie B virus, HRV= human rhinovirus.)

Virus	Coxprim 1 5′ ACCTTTGTACGCCTGTT 3′	Coxprim 2 5′ AAGCACTTCTGTTTCCC 3′	Coxprim 3 5′ AGGAGGCCGGGGACTTA 3′	Coxprim 4 5′ ATGAAACCCACAGGCAC 3′
PVI	—C—			
PV II		—C—		
PV III		—A—		
CAV 9	—C—G—	—C—		
CAV 21				
CBV 1	—G—	—A—		
CBV 3				
CBV 4	—G—	—T—G—		
HRV 1B	-A— —AT—			
HRV 2	-A— —A—			
HRV 14				
HRV 89	-A— —A—			

To test the sensitivity of the nested PCR, RNA was extracted from a series of 10-fold dilutions of a virus stock containing 5×10^6 pfu/ml PV I and used for reverse transcription and nPCR. It was possible to amplify viral RNA up to a dilution of 10^8. Considering that one pfu is usually equivalent to about 200 virus particles, the described nPCR allows the detection of less than ten viral genomes with a resulting DNA fragment in agarose gel electrophoresis (Figure 2a). To test what amount of a clinical sample is necessary for nPCR, different amounts of CSF containing PV III were tested. As figure 2b shows, it is possible to detect the virus in as little as 10µl of CSF.

For further characterisation of the amplified virus, nPCR was followed by a series of RE-digestions. The resulting RE-pattern allowed the determination of the serotype for picornaviruses with available sequence data (Table 2, Figure 3). Others could be characterised by comparison with the RE-pattern of prototype strains. Finally, direct sequencing of PCR products with the

PCR primers was done in some cases (data not shown).

To prove the suitability of the described nPCR for routine diagnostics, 134 different clinical samples have been tested. As shown in table 3, nPCR worked well with either liquid or solid samples. The CSF, serum and stool samples are from cases of aseptic meningitis (25 cases), paralytic poliomyelitis (1 case, PV III) and hand-foot-and mouth disease (1 case, CAV 21). Heart biopsies are from patients with dilatative cardiomyopathy (DCM) and obtained by cardiac catheterisation. As a control group, 70 tissue samples from patients with coronary

Table 2. Resulting DNA fragments after RE-digestion of nPCR product of picornaviral prototype strains.

Serotype	Sty I	Bgl I	Xmn I	Eco R1	Hind III	Nhe I	Mlu I	Bam H1	Nco I
PV I	71+226	297	297	297	297	297	297	58+239	71+226
PV II	102+195	297	297	297	297	297	297	297	297
PV III	297	297	297	297	297	297	105+192	297	297
CAV 9	297	80+217	61+236	297	297	29+268	297	297	297
CAV 21	83+99+114	296	296	296	296	296	296	296	83+213
CBV 1	103+194	297	61+236	297	297	297	297	297	103+194
CBV 3	10/73/214	297	61+236	297	297	297	297	297	10/73/214
CBV 4	297	80+217	61+236	297	297	297	297	297	297
HRV 1b	293	95+198	293	293	293	293	293	293	293
HRV 2	289	93+196	289	289	85+204	289	289	289	289
HRV 14	290	94+196	290	127+163	290	290	290	290	290
HRV 89	41+252	96+197	293	293	293	293	293	293	293

artery disease were obtained during open heart surgery. The finding of 6/33 (18%) positive biopsies versus 2/70 (2,8%) in non-DCM hearts, confirms the assumption that enteroviruses - particularly CBV 3 - are one of the etiological agents of DCM. (Bowels 1986, Kandolf 1988, Archard 1988, Jin 1990, Petitjean 1992). Additionally, 20 hearts of CBV3 infected mice were removed at different days post infection and tested for viral RNA. By nPCR, virus could be detected from day 1 to 14 in the infected mice, while being negative in uninfected control-mice and in infected mice at days 20 and 30 post-infection.

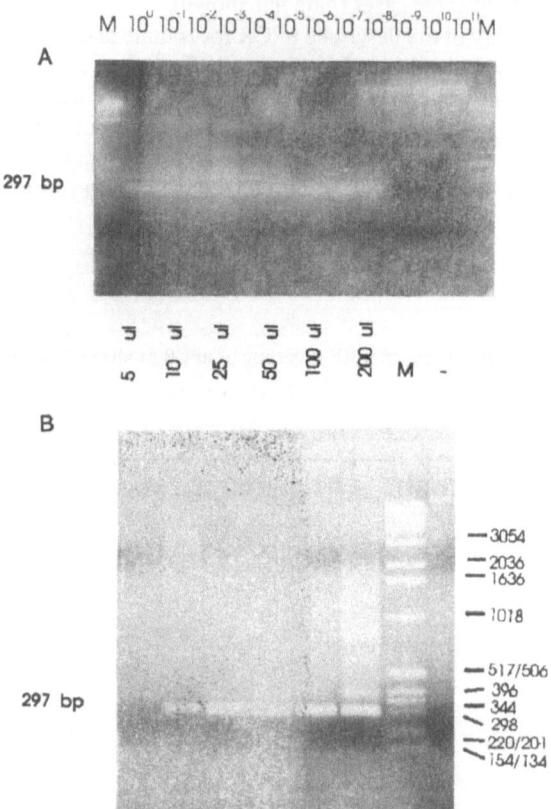

Figure 2. Determination of PCR-sensitivity: A) Agarose gel (1%) shows PCR products from supernatants of Vero-cells infected with a series of 10-fold dilutions of CBV 3 (stock solution 10^0: 5×10^6 pfu/ml). B) PCR results of different volumes (5-200 µl) of CSF of a patient with paralytic poliomyelitis caused by PV III. M: 1 kb ladder, -: CSF of a patient with HSV infection.

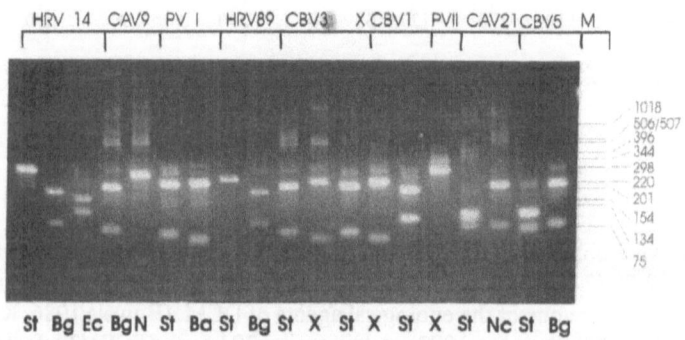

Figure 3. Agarose gel (3%) of restriction-enzyme (RE) digestion of the PCR products from nine picornaviral prototype strains and one infected mouse heart (X). St = Sty I, Bg = Bgl I, Ec = Eco RI, Nh = Nhe I, Ba = Bam HI, X = Xmn I, Nc = Nco I

Table 3. Clinical specimens tested with the nPCR for picornaviruses

Specimen	Total	PCR+	PCR -	RE-determination of serotype
CSF	17	9	8	2 x PV III 1 x CBV 3 1 x CBV 1 1 x shorter fragment like E 22 1 x CBV 5 relating RE-pattern 3 x unclear
Sera	10	2	8	1 x CBV 3 1 x CBV 5 relating RE-pattern
Feces	4	3	1	1 x CAV 21 1 x PV III 1 x shorter fragment like E 22
Heart biopsies DCM patients	33	6	27	3 x CBV 3 3 x unclear
Heart tissue Coronary artery disease patients	70	2	68	1 x CBV 4 1 x unclear

ACKNOWLEDGEMENTS

We thank the following institutions for providing enterovirus and rhinovirus strains: Landesuntersuchungsanstalt für das Gesundheitswesen, Erlangen; Institut für Virologie und Immunbiologie, Würzburg; Robert-Koch Institut, Berlin; Sandoz Forschungsinstitut, Vienna. A part of the study was supported by a grant of the Johannes and Frieda Marohn-Stiftung.

REFERENCES

Archard LC, Freeke CA, Richardson PJ, Meany B, Olsen EGJ, Morgan-Capner P., Rose ML, Taylor P, Banner NR, Yacoub MH and Bowles NE. Persistence of enterovirus RNA in dilated cardiomyopathy:A progression from myocarditis. In: Schultheiss HP (ed.), New concepts in viral heart disease, Springer-Verlag, Berlin Heidelberg 1988, 349-362

Bowels NE, Richardson PJ, Olsen EG and Archard LC. Detection of coxsackie-B-virus-specific RNA sequences in myocardial biopsy samples from patients with myocarditis and dilated cardiomyopathy. The Lancet 1986, 1120-1122

Chomczynski P. and Sacchi N. Single-step method of RNA-isolation by acid guanidinium thiocyanate-phenol-chloroform extraction. Anal. Biochem. 1987, 162:156-159

Jin O, Sole J, Butany JW, Chia WK, McLaughin PR, Lui P and Liew CC. Detection of enterovirus RNA in myocardial biopsies from patients with myocarditis and cardiomyopathy using gene amplification by polymerase chain reaction. Circulation 1990, 82:8-16

Kandolf R, Kirschner P, Ameis D, Canu A, Erdmann E, Schultheiss HP, Kemkes B and Hofschneider HP. Enteroviral heart disease: diagnosis by in situ hybridisation. In Schultheiss HP (ed), New concepts in viral heart disease, Springer Verlag, Berlin Heidelberg 1988, 337-334

Kwok S. and Higuchi R. Avoiding false positives with PCR. Nature 1989, 339:237-238

Petitjean J., Kopecka H., Freymuth F., Langlard JM, Scanu P., Galateau F., Bouhour JB., Ferriere M., Charbonneau P. and Komajda M. Detection of enteroviruses in endomyocardial biopsy by molecular approach. J. Med. Virol. 1992, 37:76-82

Rotbart H. A. Enzymatic RNA amplification of the enteroviruses. J. Clin. Microbiol. 1990, 28 :438-442

Rückert RR. Picornaviridae and their replication. In B. N. Fields, D.M. Knipe et al. (ed.), Virology, 2nd ed. Raven Press, New York 1990, 507-548

DETECTION OF HERPES SIMPLEX VIRUS USING POLYMERASE CHAIN REACTION

H.H. Kessler, K. Pierer, D. Stünzner and E.Marth

Institute of Hygiene, University of Graz, Austria

Herpes simplex virus (HSV) has a worldwide distribution. In the adult population most individuals have been exposed to HSV. In immunocompromised patients (for instance, transplant recipients or AIDS patients) both primary and recurrent infection can be life-threatening. In neonates HSV infection can have severe complications with a either fatal outcome or lasting sequelae. Therefore, the introduction of a sensitive and rapid method to detect HSV appears to be of paramount importance. The polymerase chain reaction (PCR) is a new in vitro DNA amplification technique, with high sensitivity and speed. A PCR protocol designed for detection of HSV is presented. The role of PCR in detecting HSV is reviewed and compared to other methods.

INFECTIONS PRODUCED BY HERPES SIMPLEX VIRUS

Herpes simplex virus (HSV) figures as the most important agent of sporadic encephalitis and is responsible for one of the most widespread sexually transmitted diseases (Whitley, 1990a). HSV infections are transmitted through secretions, mainly saliva or secretions from the genitourinary tract. HSV type 1 is commonly associated with oropharyngeal infections, keratoconjunctivitis, encephalitis, meningitis, myelitis, radiculitis, and cranial nerve infections, whereas HSV type 2 commonly causes genital infections. However, cases of oropharyngeal HSV 2 infections have been observed, and some primary genital infections are produced by HSV 1. Symptoms of primary infections are usually more severe than those of recurrences. Both primary and recurrent infection can be life-threatening in immunosuppressed patients (e.g. transplant recipients or AIDS patients) and in neonates. Furthermore, demonstration of HSV in specimens from patients with eye diseases and eczematous lesions has significant value.
The prognosis of severe HSV disease has improved since the introduction of modern therapeutic management. However, antiviral drugs have to be administered early. Therefore, a

Methods in DNA Amplification, Edited by
A. Rolfs *et al.*, Plenum Press, New York, 1994

sensitive and rapid method for detection of HSV is of paramount importance in order to decrease lethality and sequelae.

LABORATORY DIAGNOSIS

Viral culture is still considered the gold standard in the diagnosis of HSV against which all other diagnostic approaches should be measured (Landry and Hsuing 1986). Cell cultures with confluent monolayers of cells are inoculated with the specimen to be studied (Smith et al 1973). Usually cytopathic changes are visible within three days, but sometimes cultures require up to one week. However, demonstration of HSV by viral culture from cerebrospinal fluid is seldom successful (Nahmias et al 1982). Rapidity can be increased by centrifugation, incubating the cultures for 16 h, and using type-specific antibodies with indirect immunofluorescence (Gleaves et al 1985). The employment of shell vial culture combined with usage of type-specific monoclonal antibodies and enzyme-linked immunoassay proves rapid and specific, but lacks sensitivity compared to routine viral culture, as has been demonstrated recently (Johnston and Siegel 1990). Therefore these procedures have no clinical or laboratory advantage for HSV detection.

Serologic diagnosis is based on the demonstration of specific IgM antibodies and a significant rise in antibody titers. However, serologic studies show a delayed rise with respect to the onset of illness and are often difficult to interpret, especially in immunocompromised patients and in neonates. In a recent study a lack of correlation between virus detection and serology in patients with AIDS has been demonstrated for diagnosis of active CMV infection (Lazzarotto et al 1992), and it is strongly suggested that in these patients diagnosis should not be based on the currently available serologic tests. It can be supposed that these results are valid for all herpesviridae. Furthermore, in many patients with recurrent HSV infections serology is complicated by the difficulty in detecting significant changes in serum antibodies during the course of the disease. IgG levels might reveal only insignificant changes, and IgM antibodies may be present or absent.

Nucleic acid hybridization techniques have been applied for the direct detection of the HSV genome (Kulski and Norval 1985). However, these methods lack sensitivity (Drew 1986). Consequently, their value, mainly in the early diagnosis of HSV infections, is limited. Gene amplification using polymerase chain reaction (PCR) combined with nucleic acid hybridization results in highest sensitivity and specificity (Saiki et al 1988). Consequently, PCR has recently been introduced to detect HSV in experimental animal studies (Boerman et al 1989; Boerman et al 1992) as well as in the clinical routine for detection of cutaneous herpes simplex virus infection in adults (Cao et al 1989; Brice et al 1989) and in children (Weston et al 1992), of herpes simplex encephalitis (Rowley et al 1990; Klapper et al 1990; Aurelius et al 1991), and of genital HSV infections (Rogers et al 1992).

DETECTION OF HERPES SIMPLEX VIRUS EMPLOYING PCR

DNA extraction. The employment of an adequate DNA extraction protocol is one of the crucial points of any amplification procedure. The extraction protocol should be rapid, cheap, easy to handle and guarantee a sufficient yield of DNA. On the other hand, after extraction the template DNA should contain only a minimum of substances such as heparin and/or hemoglobine which might inhibit the DNA polymerase, and, consequently dramatically decrease sensitivity. The authors use a rapid DNA extraction protocol employing Chelex 100 resin (Bio-Rad Laboratories, California, USA). 50µl of cerebrospinal fluid (CSF) are added to 150µl of a solution consisting of 20%w/v Chelex 100 in 10mM Tris-HCl, pH 8.0, 0.1mM EDTA, and 0.1% sodium azide. After vortexing for 10 seconds the tube is incubated at 56°C for 20 minutes ; this was followed by vortexing for another 10 seconds. After incubation at 100°C for 10 minutes the tube is allowed to cool to room temperature. 20µl of the supernatant are carefully removed and used for amplification directly without further purification.

DNA amplification. To amplify HSV DNA two 22-mer oligonucleotide primers (Oligos Etc., Connecticut, USA), which flank a 92 base pair segment of the HSV 1 strain SC16 DNA polymerase gene coding region (Larder et al 1987) are used. Their sequences are as follows: 5'CATCACCGACCCGGAGAGGGAC and 5'GGGCCAGGCGCTTGTTGGTGTA. The amplified region identifies the genomes of HSV types 1 and 2, and thus the employed primers do not discriminate between these virus types (Tsurumi et al 1987). Furthermore, no cross-reaction with other herpesviruses could be found when using these primers (Cao et al 1989). Since January 1993 we have performed an additional procedure to detect the presence of HSV 1 DNA. Primers from the HSV 1 glycoprotein C gene, which amplify a region of 115 bp, are employed (Kaye et al 1991).

PCR amplification is performed in specially designed PCR tubes (Sarstedt, FRG) using a programable thermal cycler (PHC-2; Techne, UK). Thirty PCR cycles consisting of 1 min at 95°C, 1 min at 50°C and 15 sec at 72°C are run. After the final cycle, the tubes are incubated for an additional 7 min at 72°C.

Gel electrophoresis and hybridization procedure. Electrophoresis is performed on an agarose gel (3% NuSieve/1% SeaKem GTG; FMC Corporation, Maine, USA) in TBE buffer (Sambrook et al 1989) at 100 V for 1.5 h. Ten microliters of each amplified sample are electrophoresed. After staining with ethidium bromide (5µg/ml), the gel is photographed under UV light (300nm).

Hybridization is performed using the digoxigenin system (Boehringer Mannheim, FRG) according to the manufacturer's instructions. Labeling is performed by DIG-11-dUTP incorporation during PCR. The labeled DNA is hybridized to immobilized target DNA on a nylon membrane using a dot blot technique. After addition of a digoxigenin antibody-conjugate including alkaline phosphatase the hybrids are detected by usage of a chemiluminescence dye. Photographs are taken with a specially designed camera system (Analytical Luminescence Laboratory, California, USA).

EVALUATION OF THE DESCRIBED SYSTEM

Applying the procedures as described above, HSV can be detected easily from cerebrospinal fluid as well as from oral and genital swabs and peripheral blood specimens (EDTA-blood). Figure 1 shows results obtained by gel electrophoresis, figure 2 gives an example of hybridization results. Both the demand to deliver the results as quickly as possible, and the increasing probability of false-positive reactivity due to contamination in relation to the number of manipulations involved in sample processing (Clewley 1989; Kwok and Higuchi 1989) require a rapid and safe DNA extraction protocol. The authors prefer a method employing Chelex 100 resin. Chelex 100 is an analytical grade resin composed of styrene divinylbenzene copolymers containing paired iminodiacetate ions which act as chelating groups to bind polyvalent metal ions. The substrate has found many applications. These include analysis of trace metals in different fluids (Paulson 1986; Liu and Ingle Jr 1989) and removal of trace metals from fluids and solid materials (Ray and Puvathingal 1985; Laue et al 1989). Since trace metals can be concentrated by adsorption to Chelex chelating resin, preconcentration of samples for analysis has been extensively reviewed (Brady et al 1972). Recently, chelex has been successfully employed in DNA extraction protocols (Singer-Sam et al 1989; Walsh et al 1991; Jung et al 1991; Schriefer et al 1991; Gomez-Lus et al 1993). The basic Chelex procedure consists of boiling the sample in a Chelex solution, and then adding a fraction of the supernatant directly to the PCR mix. The presence of Chelex during boiling prevents the degradation of DNA by chelating metal ions that may catalyze the breakdown of DNA subjected to high temperatures in low ionic strength solutions (Singer-Sam et al 1989). However, Chelex itself inhibits PCR, mainly by disturbance of the optimum magnesium ion concentration. Therefore, to avoid transferring resin with the sample, Chelex has to settle completely before removing the extracted DNA. The authors compared the described extraction protocol to a previously reported procedure (Rowley et al 1990). Cerebrospinal fluid was taken from two patients with confirmed Herpes simplex encephalitis. One of the samples was contaminated with blood. Both methods were tested in a dilution series. Regarding the yield of DNA shown by the detection of HSV DNA after amplification, both methods gave identical results. Furthermore, these results and the applicability of the described protocol on EDTA-blood indicate strongly that haematin, which figures as inhibitor of PCR (Higuchi 1989) seems to be sufficiently removed by the resin. It must be added that we did not apply the Chelex procedure on tissue specimens yet. However, in light of a previously reported simplified extraction protocol (Cao et al 1989), it may be supposed that it also works well on such specimens.

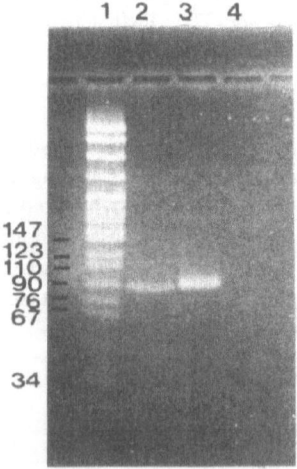

Figure 1. Agarose gel analysis of amplified HSV DNA. MspI-digested pBR322 DNA as molecular weight marker (lane 1); lane 2 patient sample (positive result); lane 3 positive control; lane 4 negative control.

The result of DNA amplification depends on the appropriate choice of the primers. The primers we employed up to the end of 1992 do not discriminate between the two types of Herpes simplex viruses. Since the therapeutical management of an infection caused by HSV 1 does not differ from that caused by HSV 2, these primers are suitable for routine clinical diagnosis. However, it is not possible to give evidence about the epidemiology of HSV 1 and HSV 2 infections. Since differences of prognosis associated with the type of HSV infection in neonates and infants have been reported (Corey et al 1988; Whitley 1990b), primers specifically designed to amplify exclusively HSV 1 or HSV 2 should be used for these patients. Furthermore, such primers are essential in experimental studies concerned with detection of HSV latency. Additionally, in pediatrics the presence of encephalitis and/or other complications of the central nervous system in the pre-eruptive stage of exanthem subitum should been taken into consideration (Yoshikawa et al 1992). Therefore, in some cases it will be necessary to employ specific primers to detect human herpesvirus 6 DNA.

The application of a hybridization system leads to an increase of sensitivity (Rogers et al 1992) and excludes non-specific amplification products (Rand and Houck 1990, Schmidt et al 1991). In contrast to gel electrophoresis the applied system allows the detection of as little as 1pg DNA. The handling of the system is easy. The detection of hybrids by enzyme immunoassay with chemiluminiscence needs only 2.5 hours.

Comparison of PCR combined with nucleic acid hybridization to other diagnostic approaches as, for instance, viral culture (Rogers et al 1992) and evidence of intrathecal synthesis of antibodies (Kimura et al 1992) have shown that PCR allows an earlier diagnosis of a suspected HSV infection.

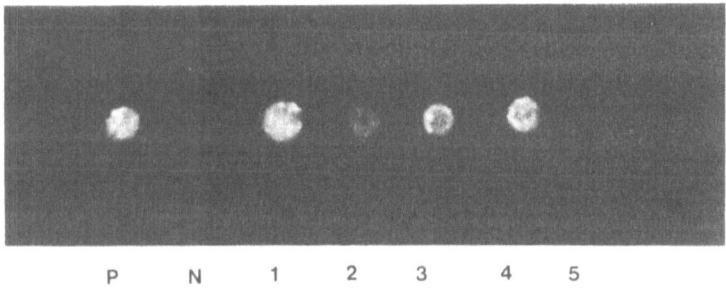

Figure 2. Dot blot of amplified HSV DNA. P positive control. N negative control. 1-5 patient samples (1-4 positive results, 5 negative result).

CONCLUSIONS AND FUTURE ASPECTS

PCR ensures the specific and highly sensitive detection of HSV. It has proved to be a valuable tool for rapid and safe detection of HSV in routine clinical applications. Application of this technique in experimental studies may lead to a better understanding of how the body controls HSV infection and how the virus maintains a latent infection. For instance, the tendency of cutaneous lesions of erythema multiforme to recur in the same location in a given patient raises some questions. Does the virus remain within the neurons innervating that cutaneous site or in the skin itself? In what form does the virus exist? What is responsible for the immunologically quiescent state? HSV has been detected in the peripheral blood of immunocompromised patients as well as in neonates with disseminated disease and in some patients during the period of active infection (Brice et al., 1992). Possible implications of this phenomenon, for instance, on the development of disseminated disease remain in discussion.

Furthermore, the role of HSV on parainfectious neurologic syndromes and paretic conditions of unknown etiology may be clearified by application of PCR.

However, this technique has some disadvantage in the monitoring of patients. PCR positivity does not necessarily correlate with clinical evidence of infection. In these cases, application of PCR using mRNA as substrate might lead to a more accurate evaluation of the activity of infection. However, techniques employing RNA remain difficult. Furthermore, since PCR results are not quantitative they do not provide a precise measure of the efficacy of antiviral treatment.

Because of these disadvantages, other methods may compete with PCR. Recently, an increased activity of a cellular transcription factor during HSV infection, when most cellular gene activity is inhibited, was described (Jang and Latchman 1992). The effect was also observed in uninfected cells transformed with a plasmid expressing a HSV immediate-early protein or in cells transfected with the gene encoding this protein. These findings may lead to the production of monoclonal antibodies and further to the development of an assay to detect immediate early HSV antigens. Such a system has been established for cytomegalovirus, and a correlation between blood levels of antigenemia and severity of cytomegalovirus disease symptoms has been shown (The et al 1990).

In summary, because of its high sensitivity and specificity, PCR combined with nucleic acid hybridization appears to be a valuable tool for detection of HSV in routine clinical applications. Additionally, the application of this method may contribute to a better understanding of the unsolved mysteries of the Herpes simplex virus.

REFERENCES

Aurelius E, Johansson B, Sköldenberg B, Staland A and Forsgen M: Rapid diagnosis of herpes simplex encephalitis by nested polymerase chain reaction assay of cerebrospinal fluid. Lancet 1991, 337: 189-92.

Boerman RH, Arnoldus EPJ, Raap AK, Bloem BR, Verhey M, van Gemert G, Peters ACB and van der Ploeg M: Polymerase chain reaction and viral culture techniques to detect HSV in small volumes of cerebrospinal fluid; an experimental mouse encephalitis study. J Virol Methods 1989, 25: 189-198.

Boerman RH, Peters ACB, Arnoldus EPJ, Raap AK, van Loon AM, Bloem BR and van der Poeg M: Polymerase chain reaction detection of herpes simplex virus in cerebrospinal fluid. In: Diagnosis of human viruses by polymerase chain rection technology, Becker Y and Darai G (eds), Springer 1992, Berlin/Heidelberg/New York.

Brady FO, Monaco ME, Forman HJ, Schutz G and Feigelson P: On the role of copper in activation of and catalysis by tryptophan-2,3-dioxygenase. J Biochem Chem 1972, 247: 7915-22.

Brice SL, Krzemien D, Weston WL and Huff JC: Detection of herpes simplex virus DNA in cutaneous lesions of erythema multiforme. J Invest Dermatol 1989, 93: 183-7.

Brice SL, Stockert SS, Jester JD, Huff JC, Bunker JD and Weston WL: Detection of herpes simplex virus DNA in the peripheral blood during acute recurrent herpes labialis. J Am Acad Dermatol 1992, 26: 594-8.

Cao M, Xiao X, Egbert B, Darragh TM and Yen TSB: Rapid detection of cutaneous herpes simplex virus infection with the polymerase chain reaction. J Invest Dermatol 1989, 92: 391-2.

Clewley JP: The polymerase chain reaction, a review of the practical limitations for human immunodeficiency virus diagnosis. J Virol Methods 1989, 25: 179-88.

Corey L, Whitley RJ, Stone EF and Mohan K: Differences between herpes simplex virus type 1 and type 2 neonatal encephalitis in neurological outcome. Lancet 1988, Vol.1: 1-4.

Drew WL: Controversies in viral diagnosis. Rev Infect Dis 1986, 8: 814-24.

Gleaves CA, Wilson DJ, Wold AD and Smith TF: Detection and serotyping of herpes simplex virus in MRC-5 cells by use of centrifugation and monoclonal antibodies 16 h postinoculation. J Clin Microbiol 1985, 21: 29-32.

Gomez-Lus, Fields BS, Benson RF, Martin WT, O`Connor SP and Black CM. Comparison of arbitarily primed polymerase chainreaction, ribotyping and monoclonal antibody analysid for subtyping *Legionella pneumophila* serogroup 1. J Clin Microbiol 1993, 31: 1940-1942.

Higuchi R: Simple and rapid preparation of samples for PCR. In: Ehrlich HA (Ed), PCR technology: principles and applications for DNA amplification. Stockton Press, USA, 1989, pp. 31-8.

Jang KL and Latchman DS: The herpes simplex virus immediate-early protein ICP27 stimulates the transcription of cellular Alu repeated sequences by increasing the activity of transcription factor TFIIIC. Biochem J 1992, 284: 667-73.

Johnston SL and Siegel CS: Comparison of enzyme immunoassay, shell vial culture, and conventional cell culture for the rapid detection of herpes simplex virus. Diagn Microbiol Infect Dis 1990, 13: 241-4.

Jung JM, Comey CT, Baer B and Budowle B: Extraction strategy for obtaining DNA from bloodstrains for DNA amplification and typing of the HLA-DQ alpha gene. Int J Legal Med 1991, 104: 145-8.

Kaye SB, Lynas C, Patterson A, Risk JM, McCarthy K, Hart CA: Evidence for herpes simplex viral latency in the human cornea. Br J Ophthalmol 1991, 75: 195-200.

Kimura H, Aso K and Kuzushima K: Relapse of herpes simplex encephalitis in children. Pediatrics 1992, 89: 891-4.

Klapper PE, Cleator GM, Dennett C and Lewis AG: Diagnosis of herpes encephalitis via southern blotting of cerebrospinal fluid DNA amplified by polymerase chain reaction. J Med Virol 1990, 32: 261-4.

Kulski JK and Norval M: Nucleic acid probes in diagnosis of viral diseases of man. Arch Virol 1985, 83: 3-15.

Landry ML and Hsuing GD: New concepts in diagnostic virology. In: Medical Virology V. Lawrence Erlbaum Associates, London 1986.

Larder BA, Kemp SD and Darby G: Related functional domains in virus DNA polymerases. The EMBO Journal 1987, 6: 169-75.

Laue TM, Lu RL, Johnson AE, Krieg UC and Esmon CT: Ca2+-dependent structural changes in bovine blood-coagulation factor VA and its subunits. Biochemistry 1989, 28: 4762-71.

Lazzarotto T, Dal Monte P, Boccuni MC, Ripalti A and Landini MP: Lack of correlation between virus detection and serologic tests for diagnosis of active cytomegalovirus infection in patients with AIDS. J Clin Microbiol 1992, 30: 1027-9.

Liu Y and Ingle Jr. JD: Automated online ion-exchange trace enrichment system with flame atomic absorption detection. Anal Chem 1989, 61: 520-4.

Nahmias AJ, Whitley RJ, Visintine AN, Takei Y, Alford Jr. CA, and the collaborative antiviral study group: Herpes simplex virus encephalitis: laboratory evaluations and their diagnostic significance. J Infect Dis 1982, 145, 829-36.

Paulson AJ: Effects of flow-rate and pretreatment on the extraction of trace-metals from estuarine and coastal seawater by Chelex 100. Anal Chem 1986, 58: 183-7

Rand KH and Houck H. Taq polymerase contains bacterial DNA of unknown origin. Mol Cell Probes 1990, 4: 146-148

Ray WJ and Puvathingal JM: A simple procedure for removing contaminating aldehydes and peroxides from aqueous solutions of polyethylene glycols and of nonionic detergents that are based on the polyoxyethylene linkage. Anal Biochem 1985, 146: 307-12.

Rogers BB, Josephson SL, Mak SK and Sweeney PJ: Polymerase chain reaction amplification of herpes simplex virus DNA from clinical samples. Obstet Gynecol 1992, 79: 464-9.

Rowley AH, Whitley RJ, Lakeman FD and Wolinsky SM: Rapid detection of herpes simplex virus DNA in cerebrospinal fluid of patients with herpes simplex virus encephalitis. Lancet 1990, 335: 440-1.

Saiki RK, Gelfand DH, Stoffel S, Scharf SJ, Higuchi R, Horn GT, Mullis KB and Erlich HA: Primer-directed enzymatic amplification of DNA with a thermostable DNA polymerase. Science 1988, 239: 487-91.

Sambrook J, Fritsch EF and Maniatis T: Molecular cloning: a laboratory manual. Cold Spring Harbor Laboratory, Cold Spring Harbor, New York, 1989

Schmidt TM, ace B and Pace NR. Detection of DNA contamination in Taq polymerase. BioTechniques 1991, 11: 176-177

Schriefer ME, Sacci Jr JB, Wirtz RA and Azad AF: Detection of polymerase chain reaction-amplified malarial DNA in infected blood and individual mosquitoes. Exp Parasitol 1991, 73: 311-6.

Singer-Sam J, Tanguay RL and Riggs AD: Use of Chelex to improve the PCR signal from a small number of cells. Amplifications 1989, 5: 11.

Smith IW, Pedutherer JF and Robertson DHH: Characterization of genital strains of herpesvirus hominis. Br J Vener Dis 1973, 49: 385-90

The TH, van der Bij W, van den Berg AP, van der Giessen M, Weits J, Sprenger HG, van Son WJ: Cytomegalovirus-antigenemia. Rev Infect Dis 1990, 12: S737-S744.

Tsurumi T, Maeno K and Nishiyama Y: Nucleotide sequence of the DNA polymerase gene of herpes simplex virus type 2 and comparison with the type 1 counterpart. Gene 1987, 52: 129-37.

Walsh PS, Metzger DA and Higuchi R: Chelex 100 as a medium for simple extraction of DNA for PCR-based typing from forensic material. Biotechniques 1991, 10: 505-13.

Weston WL, Brice SL, Jester JD, Lane AT, Stockert S and Huff JC: Herpes simplex virus in childhood erythema multiforme. Pediatrics 1992, 89, 32-4.

Whitley RJ: Viral encephalitis. N Engl J Med 1990a, 323: 242-50.

Whitley RJ: Herpes simplex virus infections. In: Remington JS and Klein JO (Ed): Infectious diseases of the fetus and newborn infant. The W.B.Saunders Co, Philadelphia 1990b, pp. 282-305.

Yoshikawa T, Nakashima T, Suga S, Asano Y, Yazaki T, Kimura H, Morishima T, Kondo K and Yamanishi K: Human herpesvirus-6 DNA in cerebrospinal fluid of a child with exanthem subitum and meningoencephalitis.

DETECTION OF HUMAN CYTOMEGALOVIRUS (HCMV) - DNA FROM GRANULOCYTES AND PERIPHERAL BLOOD MONONUCLEAR CELLS (PBMNC) OF IMMUNOSUPPRESSED PATIENTS BY NESTED PCR: LACK OF CORRELATION TO ISOLATION OF INFECTIOUS VIRUS

K. Hamprecht and H.-J. Gerth

Department of Medical Virology and Epidemiology of Viral Diseases, Hygiene-Institute, University of Tübingen, Germany

A rapid and very sensitive nested PCR method with primer oligonucleotides from the immediate early gene region of HCMV laboratory strain AD 169 was used to detect HCMV-DNA from PBMC, granulocytes, urine, CSF, bronchoalveolar lavage and autopsy samples from bone marrow or kidney transplant recipients with clinically suspected primary or recurrent CMV-infection. Results of nested DNA-PCR were compared to isolation of infectious virus by conventional cell culture and culture-immunoperoxidase staining. Prior to amplification DNA was isolated from all specimens. Native urine was inhibitory to the action of Taq polymerase. Therefore urine samples were first concentrated with PEG 6000/ NaCl. In total 60 different specimens from 20 patients were analyzed. In 43 samples HCMV-DNA could be detected. Detection rates of HCMV-DNA from PBMC (14/21) and granulocytes (11/17) were nearly identical (67% and 65% respectively). Infectious virus however could be isolated in only one case. The initial detection rate of virus isolation from leukocytes prior to ganciclovir treatment was 19% (4/21). As shown for three patients ganciclovir treatment had no influence on HCMV-DNA detection from granulocytes, PBMC and urine (within the indicated time) but detection of initially infectious virus failed after ganciclovir administration. These results clearly demonstrate that detection of HCMV-DNA from granulocytes and PBMC (containing up to 20% monocytes) from immunosuppressed patients by DNA-PCR does not correlate to viremia, which has been established as predictive indicator for HCMV disease in renal and bone marrow transplantation. In this context the persistence of HCMV-DNA in monocytes and granulocytes of asymptomatic seropositive subjects has to be discussed.

Methods in DNA Amplification, Edited by
A. Rolfs *et al.*, Plenum Press, New York, 1994

INTRODUCTION

Human cytomegalovirus is a ubiquitous herpesvirus with a seroprevalence of 50 to 90% of adults that usually causes asymptomatic infection of the immunocompetent patient. However in the fetus and immunocompromised patients after organ and bone marrow transplantation or patients with AIDS it causes severe, often life-threatening infections such as gastroenteritis, hepatitis, pneumonitis and retinitis. After primary infection HCMV undergoes latency. The source of infection in immunocompromised patients is therefore most commonly reactivation of the own persiste HCMV or reactivation of persistent virus acquired through donor organs or blood transfusions. In the population of peripheral blood mononuclear cells (PBMC) monocytes are a major site of HCMV persistence (Taylor-Wiedeman et al 1991). Recently HCMV was detected by polymerase chain reaction (PCR) in granulocytes (Boland et al 1992) and there exists evidence for initial HCMV replication in polymorphonuclear leukocytes from viremic patients (Gerna et al 1992).

In the view of the effectiveness of ganciclovir treatment of CMV-disease in immunocompromised patients it is important to make an early diagnosis of an active CMV-infection. Conventional viral culture on fibroblast monolayers is limited by the slow development (7 to 21 days) of HCMV cytopathic effects (CPE). In this study a rapid and very sensitive PCR method with nested primer oligonucleotides from the immediate early gene of HCMV (Brytting et al 1991) was used to detect HCMV-DNA in PBMC and granulocytes as well as in urine or autopsy materials from 20 immunocompromised organ or bone marrow transplant recipients with suspected primary infection or reactivation. Results from the detection of HCMV-DNA were compared to isolation of infectious virus by conventional tube cell culture.

MATERIALS AND METHODS

Clinical specimens

EDTA-blood samples (10ml), urines, and samples from bronchoalveolar lavage and cerebrospinal fluid (CSF) or autopsy samples were obtained from patients of the bone marrow and kidney transplant departments (Medizinische und CHirurgische Klinik der Universität Tübingen). In total 60 specimens from 20 patients with clinically suspected primary or recurrent HCMV-infection were analyzed.

Isolation of HCMV by conventional tube cell culture

Monolayers of MRC-5 cells or human foreskin fibroblasts (HFF) maintained in 2 ml of MEM containing 3% FCS , supplemented with penicillin (100 units/ml) and streptomycin 100 µg/ml , were grown in glass tubes and inoculated in triplicate with 0.2 ml of specimen diluted in Hanks' balanced salt solution (HBSS) containing antibiotics and amphotericin B (5 µg/ml). For specimen preparation, urine was diluted in HBSS and filtered through a sterile 0.22 µm filter unit (Millipore). Following blood cell separation the different leukocyte fractions as well as CSF samples were used directly for inoculation. Autopsy specimens (1g) were ground in HBSS with a pestle and mortar and supernatants after centrifugation were used for inoculation.

After virus adsorption (1h at 37°C) and washing fibroblast monolayers were incubated at 37°C. They were observed in duplicate for cytopathic effects (CPE) characteristic for HCMV every 3 or 4 days for up to 5 weeks.

After 48h incubation the third inoculated fibroblast monolayer tube was used for detection of HCMV immediate early antigen (IEA) expression by immunoperoxidase staining. Therefore cells were fixed in acetone for 10 min at room temperature and stained with the mouse anti-IEA (anti p72) monoclonal antibody E13 (Biosoft) for 1h at 37°C. Following incubation with a goat anti-mouse IgG antibody conjugated to horseradish peroxidase (Dako) for 1h at 37°C the insoluble reaction product was developed by adding the substrate solution of 3-amino-9-ethyl-carbazole (AEC) in acetate buffer (100 mM, pH 4.9) with diluted hydrogen peroxide at room temperature within 20 min. Brownish stained nuclei indicate HCMV-infection.

Virus strains

The laboratory HCMV strain AD 169, as well as different clinical isolates of HCMV grown on fibroblast monolayers were used for DNA-preparation of positive controls. Cell-free supernatants from monolayers exhibiting nearly 100% CPE were concentrated as outlined for urines. Specificity of DNA-amplification was controled with DNA prepared from cell cultures infected with clinical isolates of herpes simplex virus (HSV) type I or II and adenovirus. DNA prepared from uninfected fibroblast monolayers served as further negative control.

Blood cell separation procedures

To prepare leukocytes dextran T500 (6% w/v in 0.9% NaCl) was added to EDTA-blood samples (1v/10v). After sedimentation of red blood cells (30 min at room temperature) the supernatant containing leukocytes was further fractionated by Ficoll density gradient centrifugation (40 min 400g). From the resulting interphase peripheral blood mononuclear cells (PBMC) were isolated. The bottom fraction, which contained granulocytes and residual erythrocytes was resuspended in 5 ml 0.83% NH_4Cl in 10mM HEPES pH 7.0 and incubated 7 min at 37°C for lysis of erythrocytes. After several washings granulocytes were collected by centrifugation. Monocytes were isolated from PBMC by plastic adherence (1h 37°C). Plastic non-adherent cells (PNA) were further purified by nylon wool columns.

Extraction of DNA

Urine samples ranging from 3 to 20 ml were concentrated by precipitation with polyethylene glycol (PEG). NaCl and PEG 6000 (Merck) were added to respective final concentrations of 0.7 M and 8% (w/v). The mixture was vortexed, kept overnight at 4°C and then centrifuged for 30 min at 3.200g at 4°C. The supernatant was removed and the pellet was resuspended in 100-400 µl of reaction buffer for proteinase K (TEK: 10mM Tris-HCl pH 7.8, 5mM EDTA, 0.5% SDS). Proteinase K (Sigma) was added to a final concentration of 100 µg/ml .Following incubation for 45 min at 56°C proteinase K was inactivated by incubation at 95°C for 10 min. DNA was extracted once with phenol:chloroform:isoamyl alcohol (25:24:1). The organic phase was „back-extracted" with TEK. The combined aqueous phases were extracted with an (10mM Tris-HCl, 1mM EDTA, pH 7.8) equal volume of chloroform. DNA was precipitated with ethanol in the presence of sodium acetate (0.3 M, pH 5.2) and pelleted after storing at

-20°C by centrifugation at 12.000g for 15 min at 4°C. Pellets were resuspended in 30-50 µl of distilled water and amount of DNA was quantitated spectrophotometrically. 0.1-100 ng of DNA were used for nested PCR.

Blood cell preparations were resuspended in TEK with 5.000 cells per µl, followed by digestion with proteinase K and phenol/chloroform extraction as outlined for urine. *CSF* and *BAL* were extracted as blood cells .

Autopsy samples and *fibroblast monolayers* were resuspended in TE-RNAase (10 mM Tris-HCl pH 8.0, 0.1 M EDTA pH 8.0, 20 µg/ml pancreatic RNAase, 0.5% SDS) and incubated for 1h at 37°C before proteinase K was added. Proteinase K digestion was done for 3h at 50°C, followed by phenol/chloroform extraction.

Oligonucleotide primers

The primer pairs (IEP4 C/D and IEP4 A/B) used in this nested PCR-assay (reported first by Brytting et al 1991) were from the fourth exon of the IE1 gene which is located in the EcoR I-J fragment of the HCMV laboratory strain AD 169. The first part of the nested PCR assay utilized primers IEP4 C and IEP4 D to amplify a 242 bp sequence of the immediate early protein region. The second (nested) set of primers (Jiwa et al 1989) - IEP4 A and IEP4 B - amplified a 146 bp sequence.

IEP4 C : 5' TGA GGA TAA GCG GGA GAT GT   (nucleotides 1729-1748)
IEP4 D : 5' ACT GAG GCA AGT TCT GCA GT   (nucleotides 1951-1970)

IEP4 C is complementary to the antisense strand
IEP4 D is complementary to the sense strand

IEP4 A : 5' AGC TGC ATG ATG TGA GCA AG   (nucleotides 1767-1786)
IEP4 B : 5' GAA GGC TGA GTT CTT GGT AA   (nucleotides 1893-1912)

IEP4 A is complementary to the antisense strand
IEP4 B is complementary to the sense strand

PCR-assay

For single PCR either primer pair IEP4 C/D or IEP4 A/B was used. DNA-samples with up to 100 ng/5µl were amplified in 50µl final volume using a master mix solution containing (final concentrations are given) 10mM Tris-HCl pH 9.6, 50mM KCl, 0.5mM (IEP4 C/D) or 1.0mM (IEP4 A/B) of dNTPs, 0.2µg/ml BSA, 0.15µM IEP4 C/D or 0.6µM IEP4 A/B. Taq DNA polymerase (Perkin-Elmer) was omitted from the master mix solution for the first round of amplification (IEP4 C/D) in the nested PCR assay. After adding of MgCl$_2$ (10mM final concentration) and water up to the final volume the reaction mixtures for the first round were covered with 50µl of mineral oil and DNA was denatured in an initial step by incubation for 5 min at 93°C. 2.5 U Taq DNA polymerase per reaction tube were then added. The master mix solution for the second round of amplification (IEP4 A/B) contained already 2.5 U Taq polymerase. In the nested PCR assay the first round of amplification consisted of 20 cycles of denaturation (91°C x 30sec), primer annealing (56°C x 1min) and extension (72°C x 1min) followed by a final extension step for 10 min at 72°C performed in an automated thermal cycler (Omni Gene, Hybaid) . 2.5-5µl products from the first amplification were reamplified using the master mix solution IEP4 A/B in 30 incubation cycles with timing and temperatures as outlined above. Amplified DNA target molecules were analyzed by subjecting 10µl of each sample to electrophoresis (8.8 V/cm) through a 3% agarose gel (NuSieve GTG Agarose, FMC) containing 0.5µg ethidium bromide per ml in 0.5xTBE (45mM Tris-base, 45mM boric acid, 1mM EDTA pH 8.0).Gels were photographed under ultraviolet-transillumination.

RESULTS

Optimization of the DNA-amplification of various clinical specimens

Prior to amplification DNA was isolated from all investigated materials by proteinase K-treatment, phenol-chloroform extraction and precipitation with ethanol. Since PCR sample quality of granulocytes depends on the number of contaminating erythrocytes, they were lysed by NH_4Cl-treatment. The purity of the granulocyte-fraction was 97%,as revealed Pappen-heim-staining, consisting of about 80% of neutrophils (OKM 1[+]). Native urine (10% v/v) was inhibitory to the activity of Taq-polymerase. For preparation of DNA from urine samples they were concentrated first by PEG (6000) precipitation followed by proteinase K-digestion.

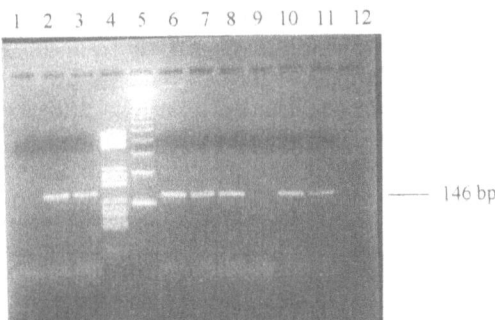

Figure 1. Detection of HCMV-DNA from PBMC and granulocytes from two transplant recipients with suspected recurrent or chronic CMV-infection by nested PCR. 100 ng of DNA isolated from each patient material were used for nested PCR. Patient A : autopsy material from a bone marrow transplant recipient with suspected CMV-pneumonitis.Patient B: blood cells and urine from a kidney transplant recipient with chronic CMV-infection. Patient C: blood cells from a viremic bone marrow transplant recipient. Size of amplified product is 146 bp. **Lanes:** 1 = 100 ng HSV I-DNA (negative control), 2 = 1 ng HCMV AD169-DNA (positive control), 3 = autopsy material from lung of patient A, 4 = pBR 322/HaeIII (size marker), 5 =·123 bp ladder (size marker), 6 = PBMC from patient B, 7 = granulocytes from patient B, 8 = urine from patient B, 9 = PBMC from patient C, 10 = granulocytes from patient C, 11 = urine from patient C, 12 = template control (no DNA). Darkish zone around 500 bp derives from loading buffer color marker (xylene cyanol FF).

Using the IEP4 C/D primer set (Brytting et al.,1991) for single PCR the lower detection limit was between 100 and 1000 fg of cell-free HCMV-DNA using laboratory strain AD 169. Sensitivity could be enhanced using the second (nested) primer set IEP4 A/B (Jiwa et al.,1989) in double PCR for detection of less than 10 fg HCMV-DNA in agarose gel electrophoresis by ethidium bromide staining without hybridization procedures, corresponding to 40 copies of HCMV-DNA. Specificity of amplification was always controlled with DNA-preparations of clinical isolates of herpes simplex virus I or II . No 146 bp bands could be detected in these controls as well as with DNA from uninfected HHF-cells in the nested PCR assay. „No template"-controls were included in each experiment. Cell-free DNA-preparations of HCMV AD169 and some clinical HCMV-isolates served as positive controls always giving the 146 bp amplification product.

Comparison of detection of HCMV-DNA by PCR with conventional virus isolation

DNA-preparations from blood cell-fractions, urine, BAL, CSF and autopsy materials from immunosuppressed patients after bone marrow or kidney transplantation were used for amplification. In total 60 samples from 20 patients with clinically suspected (mostly) recurrent CMV-infection were examined. In 43 of these samples HCMV-DNA could be detected, but in only 5 cases infectious virus could be isolated from the same sample sources (Tab.I). Detection rates of HCMV-DNA from PBMC (14/21) and granulocytes (11/17) were nearly identical (67% and 65% respectively). However infectious virus could only be isolated in one case. These results were obtained for the most part with materials from patients under or after ganciclovir treatment. Initial detection rate for isolation of infectious virus from leucocytes was 19% (4/21). Virus isolation was more sensitive than culture-immunoperoxidase (Tab.I). Ganciclovir treatment of patients has no influence on detection of HCMV-DNA from PBMC, granulocytes or urine by nested PCR up to 4 weeks after treatment. The isolation of infectious virus however is strongly influenced by former ganciclovir treatment . As shown in Tab.II for three patients from whom initially infectious virus could be detected in cell culture, after ganciclovir treatment that was not possible. But from two further patients virus could be isolated up to 2 months after ganciclovir administration.

Table 1. Detection of HCVM-DNA from granulocytes, peripheral blood mononuclear cells (PBMC) and other clinical specimens from transplant recipients with suspected primary or recurrent HCMV-infection by nested PCR in comparison to culture-immunoperoxidase-staining (CIP) and isolation of infections virus by conventional tube cell culture. [1]Number positive/total number examined. [2]Culture-immunoperoxidase staining with anti-IEA mAB (anti p72). [3]Virus isolation from the same specimen source before initiation of ganciclovir-treatment. For this monitoring of viremia the whole leukocyte fraction was used.

Specimen	Sample number	HCMV-DNA[1]	CIP[2]	virus isolation[3]	former isolation
PBMC	21	14/21	1/21	1/21	4/21
Granulocytes	17	11/17	0/17	0/17	
Urine	13	11/13	0/13	3/13	3/13
BAL	3	2/3	0/3	0/3	2/3
CSF	2	1/2	0/2	0/2	0/2
Autopsy	4	4/4	1/4	1/4	-
in total: **20 patients**	60	43/60	2/60	5/60	

Detection of HCMV-DNA in leukocyte subfractions

By blood cell-fractionation HCMV-DNA could be isolated both from PBMC and granulocytes (Table 1). In the PBMC-population up to 20% of cells are monocytes. But in some cases HCMV-DNA could be detected either in granulocytes (patient C in figure1) or in monocytes only (data not shown). In these cases it was not possible to detect HCMV-DNA in DNA-preparations from non-adherent cells (PNA) consisting of 80% CD3+-T-cells and 15% CD56+-NK-cells.

Table 2. Influence of ganciclovir (9- [1,3-dihydroxy-2-propoxymethyl]-guanine)-treatment of transplant recipients with suspected recurrent HCMV-ionfection on detection of HCMV-DNA, viremia and viruria. [1]BMTR: bone marrow transplant recipient; KTR: kidney transplant recipient. [2]Detected by PCR. [3]Incubation time before positive staining is given (days). [4]Indicated time delay (days) up to registration of HCMV-specific CPE. [5]Patients A and C: 2 weeks later, patient B: 4 weeks later. [6]not determined. += positive result, -: negative result

Patient[1]	specimen	before treatment			after treatment[5]		
		HCMV-DNA[2]	CIP[3]	virus isolation[4]	HCMV-DNA	CIP	virus isolation
A, BMTR	leukocytes	n.d.[6]	+ (2d)	+(31d)	+	-	-
	PBMC	+	-	-	+	-	-
	granulocytes	+	-	-	+	-	-
B, BMTR	leukocytes	n.d.	+ (1d)	+ (17d)	n.d.	-	-
	PBMC	n.d.	n.d.	n.d.	+	-	-
	granulocytes	n.d.	n.d.	n.d.	+	-	-
C, KTR	urine	+	-	+ (20d)	+	-	-
	PBMC	n.d.	n.d.	n.d.	+	-	-
	granulocytes	n.d.	n.d.	n.d.	+	-	-

DISCUSSION

In this study a very sensitive PCR-assay using nested primer oligonucleotides from the fourth exon of the HCMV immediate early gene located in the EcoRI J fragment of laboratory strain AD 169 (Brytting et al 1991 and Jiwa et al 1989) was used to amplify DNA-preparations from PBMC, granulocytes, urine and other clinical specimens from immunosuppressed patients with suspected primary or recurrent HCMV-infection. This assay detects as few as 10 fg HCMV-DNA. It has been well established, that HCMV viremia demonstrated by conventional virus isolation is predictive of HCMV disease in recipients of allogenic bone marrow transplant (Schmidt et al.,1991), whereas detection of viral DNA in blood leukocytes by PCR is not clinically significant in the absence of viremia or antigenemia (Gerna et al 1991). The presented results from this study show clearly that very sensitive detection of HCMV-DNA in granulocytes and PBMC (containing up to 20% monocytes) cannot be correlated to viremia by isolation of infectious virus in conventional cell culture. In this context the following aspects have to be considered. First HCMV-DNA can be present in granulocytes of immunosuppressed patients without further evidence of active CMV-infection (Boland et al1992) as well as in monocytes from healthy seropositive subjects. Monocytes are a major site of HCMV persistence in PBMC (Taylor-Wiedeman et al 1991) . It has been suggested that the monocyte/macrophage serves as a source of viral amplification and dissemination (Ibanez et al 1991). Recently, evidence was found for initial replication of HCMV in polymorphonuclear leukocytes from viremic patients (Gerna et al 1992). In a first set of experiments we were not able to detect HCMV-DNA from healthy seropositive blood donors in DNA-preparations from leukocyte subfractions (Hamprecht et al). Neither HCMV-RNA nor infectious virus could be detected in a follow-up from leukocytes of immunocompromised patients, who had detectable HCMV-DNA in these cells (Ratnamohan et al 1992). Secondly ganciclovir treatment of patients (with low virus burden) seems to influence the isolation of infectious virus in conventional cell culture (Table 2).

Taken together the detection of HCMV-DNA in leukocytes by PCR as a predictive indicator for HCMV disease and for control of ganciclovir treatment remains uncertain. RT-PCR for detection of HCMV-RNA may help to answer open questions about detection of non-viable virus or persistent HCMV-DNA by DNA-PCR.

REFERENCES

Boland GJ, De Weger RA, Tilanus MGJ, Ververs C, Bosboom-Kalsbeeek K and De Gast GC. Detection of cytomegalovirus (CMV) in granulocytes by polymerase chain reaction compared with the CMV antigen test. J. Clin. Microbiol. 1992, 30: 1763-67

Brytting M, Sundqvist V-A, Stalhandske P, Linde A and Wahren B. Cytomegalovirus DNA detection of an immediate early protein gene with nested primer oligonucleotides. J. Virol. Methods 1991, 32: 127-38

Gerna G, Zipeto D, Parea M, Revello MG, Silini E, Percivalle M, Zavattoni M and Milanesi G. Monitoring of human cytomegalovirus infections and ganciclovir treatment in heart transplant recipients by determination of viremia, antigenemia and DNAemia. J. Infect. Dis. 1991, 164: 488-98

Gerna G, Zipeto D, Percivalle E, Parea M, Revello MG, Maccario R, Peri G and Milanesi G. Human cytomegalovirus infection of the major leukocyte subpopulations and evidence for initial viral replication in polymorphonuclear leukocytes from viremic patients. J. Infect. Dis. 1992, 166: 1236-44

Hamprecht K, Sorg G, Steinmaßl M, and Gerth H-J and Jahn G. submitted

Ibanez CE, Schrier R, Ghazal P, Wiley C and Nelson JA.Human cytomegalovirus productively infects primary differentiated macrophages. J. Virol. 1991, 65: 6581-88

Jiwa NM, Van Gemert GW, Raap AK, Van de Rijke FM, Mulder A, Lens PF, Salimans MMM, Zwaan FE, Van Dorp W and Van der Ploeg M. Rapid detection of human cytomegalovirus DNA in peripheral blood leukocytes of viremic transplant recipients by the polymerase chain reaction. Transplantation 1989, 48: 72-76

Ratnamohan VM, Mathys JM, McKenzie A and Cunnigham AL. HCMV-DNA is detected more frequently than infectious virus in blood leucocytes of immunocompromised patients: a direct comparison of culture-immunofluorescence and PCR for detection of HCMV in clinical specimens. J. Med. Virol. 1992, 38: 252-59

Schmidt GM, Horaks DA, Niland JC, Duncan SR, Forman SJ and Zaia JA. A randomised, controlled trial of prophylactic ganciclovir for cytomegalovirus pulmonary infection in recipients of allogenic bone marrow transplant. N. Engl. J. Med. 1991,324:1005-11

Taylor-Wiedeman J, Sissons JGP, Borysiewicz LK and Sinclair JH. Monocytes are a major site of persistence of human cytomegalovirus in peripheral blood mononuclear cells. J. Gen. Virol. 1991, 72: 2059-64

TIMING OF MOTHER-TO CHILD TRANSMISSION OF HIV-1 AS DETERMINATED BY NESTED POLYMERASE CHAIN REACTION - A COHORT STUDY IN KIGALI, RWANDA

A. Simonon[1], Ph. Lepage[2], E. Karita[1], D.-G. Hitimana[2], F. Dabis[3], Ph. Msellati[3], F. Nsemgumuremyi[1], Ch. van Goethem[2], A. Bazubagira[2] and Ph. van de Perre[1]

[1]National AIDS Control Programme, Aids Reference Laboratory, Kigali, Rwanda, [2]Department of Paediatrics, Centre Hospitalier de Kigali, Rwanda, [3]INSERM U330, University of Bordeaux II, France

The relative contributions of in utero, intrapartum and postnatal transmission routes of HIV-1 were estimated by PCR' testing of 218 infants and children born to seropositive mothers, in Kigali, Rwanda. The study population consisted of 47 infected children, 139 uninfected children and 32 with indeterminate status. The infection status was established according to the clinical and serological profiles after a follow-up of 24 months (Ghent 1992). Nested PCR (two rounds of amplification) were performed on PBMCs isolated from serial blood samples. After a follow-up of two years, the estimated mother-to-child transmission rate of HIV-1 is 25.3%. Minimal and maximal estimates of in utero plus intrapartum transmission rate are 7.6% to 17.2% and of postpartum transmission rate 8.0% to 17.7%. The present study confirms that HIV-1 can be transmitted in the postnatal period through breastfeeding.

INTRODUCTION

In 80% of paediatric cases of Acquired Immunodeficiency Syndrome (AIDS), the Human Immunodeficiency Virus type 1 (HIV-1) is transmitted from Mother to Child (Pizzo and Wilfert 1989). Three major routes of transmission are possible: in utero, during labor or delivery, or in the postpartum period (Pizzo 1990, Douglas and King 1992, van de Perre et al 1992, Dunn et al 1992). Until now, the relative contribution of each modes of transmission remains unclear. The recognition of the mechanisms of transmission is however important to initiate prophylatic and antiretroviral treatment. The purpose of this study is to estimate the timing of Mother-to-Child transmission of HIV-1 in a cohort of 218 breastfed children born to

Methods in DNA Amplification, Edited by
A. Rolfs *et al.,* Plenum Press, New York, 1994

HIV-1 infected mothers, followed during two years, by means of a nested polymerase chain reaction (PCR) in Kigali, Rwanda.

STUDY POPULATION

Since November 1988, a prospective cohort study of the perinatal transmission of HIV-1 has been ongoing in Kigali Hospital, in Rwanda. Details of the enrolement procedures and the results of the first two years of the follow up of this cohort have already published by Lepage et al (1991).

The present study population consisted of 218 breast fed infants and children born to HIV-1 infected mothers. The mean time of breastfeeding was 560 days (STD 332days).

According to serological and clinical criteria developed by an international working group on Mother-to-Child transmission of HIV (European Economic Community AIDS Task Force 1992, Lepage et al 1993, WHO 1986), 47 children were judged to be infected (group 1), 139 uninfected (group 2) and in 32 the status remains indeterminate (group 3) after a follow-up of 24 months. Table 1 summarizes the first two years of follow up of the children born to HIV-1 infected mothers. One hundred and three uninfected children born to HIV-1 seronegative mothers (n=215) were blindly selected. In addition, a random sample of 31 HIV-1 seropositive mothers and of 13 HIV-1 seronegative mothers were tested as positive and negative internal controls, respectively.

Table 1 . Summary of the first two years of follow up of the children (n=218) born to HIV-1 infected mothers and of samples available for PCR at each time of the follow up.*PCR on 13-24 months samples was not systematically performed but was restricted to HIV-1 infected specimens with a previous negative PCR result.

Time after birth (months)	Cumulative number of	Cumulative number of lost to follow up	Number of children followed	Number of samples available for PCR
0	-	-	218	185
3	11	-	207	151
12	27	12	179	136
24	40	21	157	11*

SEROLOGICAL METHODS

Sera from cord and peripheral blood were collected every three months from mothers and children and screened for the detection of HIV-1 antibodies by a commercial Immuno Enzymatic Assay (EIA, Vironostika, Organon Teknika, Boxtel, the Netherlands). All positive samples were confirmed by a commercial Western blot (Du Pont de Nemours, Wilmington, Del, USA).

NESTED POLYMERASE CHAIN REACTION (PCR)

Two rounds of amplification were performed, as previously described (van de Perre et al 1991), on peripheral blood mononuclear cells (PBMCs) isolated from blood collected at birth (cord blood), at 3, 6-12 and 18-24 months of age.

PBMCs were separated over a 5.6 percent Ficoll gradient and 9.6 percent of sodium metrizoate (Lymphoprep, Nycomed AS, Oslo, Norway) and stored at -70°C. The cell pellets were treated with 0.2 mg of proteinase K per mililiter, followed by extraction with phenol and chloroform before precipitation in absolute ethanol.

In the first amplification, 1μg of target DNA was mixed with 50 μl of reaction buffer (100mM Tris-hydrochloric acid [pH8.3], 500mM potassium chloride and 1.5mM magnesium chloride), 1μM of primers (gag881-882, pol 001-004 and env 401-404), 200 μM of each of the deoxynucleoside triphophates (dATP, dCTP, dTTP, dGTP) and 2U of Taq polymerase (Perkin-Elmer Cetus, Norwalk, Conn.). For each primer used, the buffer concentration of magnesium chloride had to be optimized. DNA amplification was carried out by 40 cycles (denaturation 94°C for 36 sec, annealing 50°C for 42 sec and extension 68°C for 3 min.) in a Perkin Elmer Cetus DNA Thermal cycler. In the second round, 1 μl of the first reaction mixture was amplified in a total volume of 20 μl using the same programme as described above with 3 pairs of primers located within the region amplified initially (gag 883-990, pol 002-003 and env 402-403). The second round PCR mixture was analyzed by electrophoresis on a 4 percent low-melting-point agarose gel (Seakem GTG, 1 percent, and Nusieve GTG, 3 percent) and visualized by ultraviolet fluorescence. A PCR test was considered positive when there was a detectable signal for at least two of the three primer pairs.

All lysates with a negative PCR result on the cord blood were amplified for the conserved region of the human leucocyte antigen with Gh26-Gh27 primers (Saiki et al. 1986). Samples with a negative PCR result for HLA-DQ alpha gene were excluded from the analysis.

STATISTICAL ANALYSIS

The Mann-Whitney test was used for comparison of the median number of PCR tests performed in different groups of subjects (European Economic Community AIDS Task Force, 1992).

The Kaplan-Meier method was used to calculate the probability of having a positive PCR according to age at the time of testing in the group of HIV-1 infected children born to infected mothers.

RESULTS

After a follow-up of two years (Table 1), 482 PCR tests had been performed in children born to HIV-1 infected mothers and 131 in those born to uninfected mothers. All the children (47/47) of group 1 presented with a positive PCR on at least one occasion. Thirty percent of cord blood samples of these children and 68% at 3 months of age were PCR positive.

Eleven positive PCR were detected in the 139 HIV-1 uninfected children born to infected mothers (group 2). Six of them were cord blood samples and the other five were collected at 3 and at 9 months of age. All of them were tested further and were PCR negative.

These PCR positive results observed on one occasion in cord blood samples suggest a maternal blood cell contamination of cord blood specimens. The other five false positive results performed on peripheral blood samples could be due to laboratory errors or a mix-up. Six out of the 32 children with indeterminate status (group 3) had at least one positive result. This group had a lower number of blood samples available for testing than group 1 and 2 mainly because of being lost to follow-up. In uninfected infants born to seronegative mothers (group 4), DNA was detected in only one child in cord blood and 24 month samples (Table 2).

In the present study, the sensitivity of our nested-PCR was 100% (47/47). The specificity was estimated at 98% (335/342) (95% confidence interval: 96.5-99.5 %) if the results obtained in cord blood samples are excluded.

Our present data allow an estimate of the timing of mother-to child transmission. According to the hypothesis that all the children from the group 1 with a positive PCR at 3 months of age had been infected in utero or intrapartum,the minimal postnatal transmission rate was estimated at 8% (25.3% x 32%/100) and the maximal in utero plus intrapartum transmission rate is estimated at 17.2% (25.3% x 68%/100). According to the second hypothesis, that all the infected children with a negative PCR on cord blood had been infected during postpartum period, the maximal postnatal transmission rate is 17.7% (25.3% x 70%/100) and the minimal in utero plus intrapartum transmission rate is 7.6% (25.3% x 30%/100).

Table 2 . Proportion of children with a positive PCR test on at least one occasion of the follow-up according to their HIV-1 infection status at 24 months of age, Kigali (Rwanda), 1988-1991. * 6 positive PCR results obtained on cord blood samples suggesting maternal blood contamination, ** 5 specimens (group 2) and 3 (group 4) with positive PCR result on one occasion, NA samples not available by definition.

Months	Proportion of Positive PCR by age at blood collection				Children with min. one positive PCR	Median number of blood samples available for PCR per child
Groups	Birth	3	6-12	13-24		
I n=47	14/42	25/35	22/27	8/8	47/47 (100%)	3(1-5)
II n=139	6/117	3/102	2/106	0/3	11*/139(8%)	3(1-4)
III n=32	5/26	5/14	0/2	NA	16/32(31%)	2(1-3)
IV n=103	3/95	0/21	0/11	1/4	3**/131(1%)	1(1-4)

CONCLUSIONS

This study provides direct information on the timing of mother-to-child transmission of HIV-1 in Africa. It shows that our nested-PCR assay is extremely sensitive (100%) and highly specific (98%) and can be successfully used for such studies in developing countries where the current PCR assays using radioisotopes is not readily accessible.

Viral DNA was not detected in the majority of our tested cases on cord blood samples (70%) from HIV-1 infected children. This is probably due to the low amount of circulating virus. Therefore, the number of viral copies could then fall below the detectable level of the PCR assays. An active replication of HIV-1 could occur during the first week of life. Alternatively, transmission of HIV-1 could occur at the end of pregnancy or at delivery.

A maternal blood cell contamination of 6 out of 139 cord blood samples was observed, suggesting that cord blood is probably not a suitable material for early diagnosis of HIV-1 infection as already described (Ehrnst et al 1991).

Due to the uncertain reliability of the cord blood (thirty percent of positive PCR), the interpretation of the PCR results remains problematic during the neonatal period. This has important consequences for the analysis of our data.

As proposed by the Pediatric Virology Committee of the Aids Clinical Trial Group (Bryson et al 1992), all infected children who had a negative PCR at birth should be considered as having been infected either during delivery or postnatally via breastfeeding. Using this definition, the minimal estimate of in utero transmission rate would be 7.6% and the maximum combined intrapartum and postpartum transmission rate 17.7%. If only children with a positive PCR obtained after 3 months of age were indeed infected in the postnatal period, the maximum estimate of the combined in utero and in intrapartum transmission rate would be 17.2% and the minimal estimate of the postnatal transmission rate 8%.

Our present estimation of postnatal transmission rate range from 8 to 17.7% supports our recent observations on seroconverted lactating mothers (van de Perre et al 1991). It also confirms the hypothesis that breastfeeding is at least partly responsible for the excess of mother-to-child transmission rate observed in populations where breastfeeding is the rule as compared with industrialised countries where breastfeeding is disencouraged in HIV-1 infected mothers (Report of consensus workshop 1992, Ruff et al 1992, WHO 1992).

ACKNOWLEDGEMENT

We are indebted to the Minister of Health of the Rwandan Republic, Dr.C.Bizimungu and Dr.J.B.Butera (National Aids Control Programme, Kigali, Rwanda) for their support in this study; to Dr. L.Fransen (EEC - Aids Task Force, Brussels, Belgium), Mr.L.Nunes de Carvolho (EEC Delegation in Kigali) and Dr. J. F. Ruppol (Belgian Ambassy in Kigali) for their collaboration; to Prof. J. P.Butzler, Prof. A. Burny, Prof. C. Peckham, Prof. R. Salamon, Dr. E. Fox and Dr. J. Ladner for constant encouragement and advice. We thank Dr. D. Sondag-Thull and Ms D.Vaira (Transfusion Center, University of Liege, Belgium) for performing the quality control of our PCR Laboratory, and the paramedical-staff members of the mother-to child transmission study for their active collaboration and the nurses of Pediatrics Department of the Centre Hospitalier de Kigali for their devoted and compassionate care of patients with AIDS.

REFERENCES

Bryson YJ, Luzuriaga K, Sullivan JL, Wara DW. Proposed definitions for in Utero versus intrapartum transmission of HIV-1. N Engl J Med 1992, 327 : 1246-1247.

Burgard M, Mayaux MJ, Blanche S, et al. The use of viral culture and p24 antigen testing to diagnose human immunodeficiency virus infection in neonates. N Engl J Med 1992, 327 : 1192-1197.

Courgnaud V, Laure F, et al. Frequent and early in Utero HIV-1 infection. Aids Research and Human Retrovirus 1991, 7 : 337-741.

Douglas GC, King BF. Maternal-fetal transmission of human immunodeficiency virus : a review of possible routes and cellular mechanisms of infection. Clin Infect Dis 1992, 15 : 678-691.

Dunn DT, Newell ML, Aedes AE and Peckham C. Risk of human immunodeficiency virus type 1 transmission through breastfeeding.Lancet 1992, 340: 585-588.

Ehrnst A, Lindgren S, Dictor M, et al. HIV in pregnant women and their offspring ; evidence for late transmission. Lancet 1991, 338, 27 : 203-207.

European Economic Community AIDS Task Force (EEC-ATF). Workshop on Mother-to-Child transmission of HIV. Ghent, Belgium, 17-20 February 1992. Final report. EEC-ATF, Brussels, 1992 : 56p.

Krivine A, Firtion G, Coa L, Francouval C, Henrion R, Lebon P. HIV replication during the fisrt weeks of life. Lancet 1992, 339 : 1187-1189.

Lepage P, Dabis F, Hitimana DG et al. Perinatal transmission of HIV-1: lack of impact of maternal HIV infection on characteristics of livebirths and on neonatal morality in Kigali, Rwanda. AIDS 1991, 5: 295-300.

Lepage P, Van de Perre P, Msellati P et al. Mother-to-child transmission of human immunodeficiency virus type 1 (HIV-1) and its determinants : a cohort study in Kigali, Rwanda. Am J Epidemiol 1993, in press.

Pizzo PA and Wilfert CM - Pediatrics Aids, Edts Williams&Wilkins : 813pp.

Pizzo PA . Pediatrics Aids : Problems within problems. J Infect Dis 1990, 161: 316-324.

Report of consensus workshop, Sienna, Italy, January 17-18, 1992, 5 : 1019-1020.

Ruff et al. Breastfeeding and maternal-infant transmission of Human immunodeficiency virus type 1. J Pediatrics 1992 : 325-328.

Saiki RK, Bugawan TL, Horn GT, Mullis KB, Ehrlich HA. Analysis of enzymatically amplified beta-globin and HLA-DQ alpha DNA with allele-specific oligonucleotide probes. Nature 1986, 324: 163-66.

Tudor-Williams G. Early diagnosis of vertically acquired HIV-1 infection. AIDS 1991, 5 : 103-105.

Van de Perre P, Lepage P, Homsy J and Dabis F. Mother-to-Infant transmission of HIV-1 by breast milk : presumed innocent or presumed guilty? Clin Infect Dis 1992, 15: 502-507.

Van de Perre P, Simonon A, Msellati P, et al. Postnatal transmission of human immunodeficiency virus type 1 from mother-to-infant. A cohort study. N Engl J Med 1991, 325 :593-598.

World Health Organization. Acquired immunodeficiency syndrome (AIDS) Wkly Epidemiol Rep 1986, 61: 69-73.

World Health Organization, Global Programme on Aids. Consensus statement from the WHO/ Unicef consultation on HIV transmission and breast feeding . Weekly Epid Rep 1992, 24 : 177-179.

ONE-DAY DETECTION OF ENTEROVIRUSES IN CLINICAL SPECIMENS BY MAGNETIC BEAD EXTRACTION OF VIRAL RNA AND NESTED POLYMERASE CHAIN REACTION

P. Muir, F. Nicholson and J.E. Banatvala

Department of Virology, United Medical and Dental Schools of Guy's & St Thomas' Hospitals, (St Thomas' Campus), London, UK

INTRODUCTION

The enteroviruses are a large group of positive strand RNA viruses belonging to the *Picornaviridae* family, At least 70 serotypes are known to cause infection in humans, including the polioviruses, group A and B coxsackieviruses and the ECHO viruses. Viral replication occurs initially in mucosal tissue of the gastrointestinal tract. Following viraemic spread, secondary replication may occur in various target organs, including cardiac or skeletal muscle, the central nervous system, the skin or mucous membranes. Enterovirus infection is probably mild or asymptomatic in most cases, but may be associated with acute myopericarditis, paralytic poliomyelitis, aseptic meningitis, or, more rarely, encephalitis (reviewed by Melnick 1990). Laboratory diagnosis is necessary to confirm the clinical diagnosis and to distinguish enterovirus-related disease from similar conditions of non-enteroviral aetiology. However this is not always straightforward. Athough enterovirus may be detected by culture in cerebrospinal fluid (CSF), stool or throat swabs, virus is rarely cultured from patients with cardiac or skeletal muscle symptoms, which usually occur as a post-acute manifestation of infection. Furthermore, some enterovirus serotypes grow poorly in cell culture. Serological diagnosis is complicated by the large numbers of serotypes, the difficulty in demonstrating significant rises in antibody titre in many patients, particularly those with post-acute symptoms, and the high prevalence of virus-specific antibody, reflecting past exposure, in the general population.

There is significant genetic homology between different enterovirus serotypes, and this has facilitated molecular detection of enterovirus RNA by hybridization using complementary DNA or RNA probes derived from conserved regions of the genome, or by polymerase chain reaction (PCR) using primers derived from highly conserved sequences (reviewed by Rotbart

Methods in DNA Amplification, Edited by
A. Rolfs *et al.*, Plenum Press, New York, 1994

1991; Muir and Kandolf 1993). However hybridization techniques are too complex for routine clinical diagnosis, and the application of PCR for diagnosis for RNA virus infections has been limited by the laborious nature of most RNA extraction methods.

We have used magnetic bead technology to develop a rapid and simple method of extracting enterovirus RNA from clinical samples (Muir et al 1993). This is based on the selective hybridization of enterovirus RNA to a specific oligonucleotide bound to a magnetic bead solid phase, and requires 10-15 minutes per specimen. This procedure was found to be of similar sensitivity to RNA extraction methods which involve organic solvent extraction and ethanol precipitation of RNA. Extracted RNA was amplified by reverse transcription and PCR, followed by Southern blot hybridization. Magnetic bead extraction and PCR was more sensitive than isolation in tissue culture for the detection of enteroviruses in clinical specimens, and results could be obtained within two days, if required. We have now improved our extraction protocol and reduced the time required yet further by using a nested PCR format (Nicholson et al 1994). The use of nested PCR has been found to give highly specific detection of target sequences without the need for Southern blot confirmation. This modification further reduces the time required to complete the analysis, and results can be obtained within one day if required.

METHODS

Full technical details of the enterovirus detection protocol are given elsewhere (Muir et al 1993; Nicholson et al 1994). Enterovirus RNA extraction from clinical specimens or infected cell culture involved an initial treatment of sample with extraction buffer containing the chaotropic agent guanidine thiocyanate to release viral RNA from virus particles or infected cells, and to inactivate endogenous RNases present in the sample. Liquid specimens, such as cerebrospinal fluid (CSF) throat washings, stool filtrate, urine, serum/plasma, whole blood or peripheral lymphocytes, and cultures cells or culture supernatant were simply mixed with extraction buffer. Cryopreserved solid tissue was initially ground to powder in liquid nitrogen, then homogenised in extraction buffer, and sections of formalin-fixed tissue were deparaffinized then boiled in 0.1% SDS for 5 minutes prior to addition of extraction buffer.

Without further treatment, enterovirus RNA was then isolated from the specimen by selectively hybridizing to magnetic beads coated with an oligonucleotide which is complementary to a region within the target sequence to be amplified. This sequence is common to those enterovirus serotypes known to infect humans that have been sequenced to date. The 3 prime end of the enterovirus genome is polyadenylated, and it is also possible to use oligo d(T)-linked magnetic beads for enterovirus RNA extraction. However, because the target sequence to be amplified is located within the highly conserved 5' non-translated region of the genome, this would require isolation of intact viral RNA molecules of over 7000 bases long. This is feasible in many cases, but is unlikely to be successful when using stored specimens, particularly archival, formalin-fixed tissue.

Following hybridization of viral RNA to magnetic beads, the beads were washed to remove non-specific components, and used as template for reverse transcription and nested PCR. This procedure is considerably faster, and more technically straightforward than any other RNA extraction protocol. Elution of target RNA from magnetic beads is not necessary, unless long-term storage of RNA in 70% ethanol is envisaged.

Primers for reverse transcription and nested were selected to hybridize to highly conserved regions of the enterovirus genome within the 5 prime non-translated region (Muir et al., 1993; Nicholson et al 1994). Primers were selected to amplify short regions of viral RNA (228 bases for first round PCR; 148 bases for nested PCR) to minimise loss of sensitivity when studying archival samples where RNA may be fragmented due to partial degradation. The boiling procedure used to release nucleic acids from formalin-fixed tissue may also result in fragmentation of RNA. Measures to eliminate PCR contamination were employed as described previously (Muir et al 1993).

Table 1. Results of nested PCR and virus isolation in cell culture for detection of enteroviruses in clinical specimens. Patient numbers correspond to position numbers in Figure 1.

Sample No.	Clinical details (age)	Specimen	PCR result	Culture result
1	Aseptic meningitis (37)	CSF	+	not tested
2	Meningism (33)	Stool	+	Echovirus 11
3	Healthy adult	Stool	-	-
4	Polymyositis (38)	Stool	+	-
5	Acute pericarditis (56)	Stool	+	-
6	" "	Pericardial fuid	+	Enterovirus
7	Malaise (59)	PBL	+	Toxic
8	"	Stool	+	-
9	Healthy adult	Stool	-	-
10	Bornhom disease (37)	Stool	+	-
11	Polio vaccinee (5/12)	Stool	+	Poliovirus 2

RESULTS AND DISCUSSION

This enterovirus nested PCR has been successfully used to amplify RNA extracted from numerous reference strains and clinical enterovirus isolates of known serotype, including poliovirus types 1-3, coxsackievirus types A9, A16, B1-B6, echovirus types 4, 6, 7, 11, 18, 30, and enterovirus type 70. In titration experiments using enterovirus-infected cell culture supernatant, this detection protocol was found to be capable of detecting 0.1 TCID$_{50}$. Representative results of studies on clinical specimens are shown in figure 1 and table 1. From this it can be seen that nested PCR detection is capable of detecting clinical enterovirus isolates directly in clinical specimens, including poliovirus of presumed vaccine origin, is more sensitive than cell culture detection, and is useful for detecting enteroviruses in specimens which are toxic in cell culture.

We anticipate that this enterovirus detection protocol may have a number of uses, as follows:

a) Rapid diagnosis of acute enterovirus infection. The ability to detect enteroviruses in clinical specimens within one day represents a major advance in enterovirus diagnosis. This may greatly assist patient management, particularly in those with encephalitis where it is important to exclude a herpes simplex virus aetiology, and in patients with acute myocarditis, where signs and symptoms may be similar to those of acute myocardial infarction. Virus serotypes which grow poorly in culture may also be detected by PCR. We have also detected enterovirus which could not be cultured in stool samples from patients with cardiac and skeletal muscle disease (Muir et al 1993). This requires further evaluation.

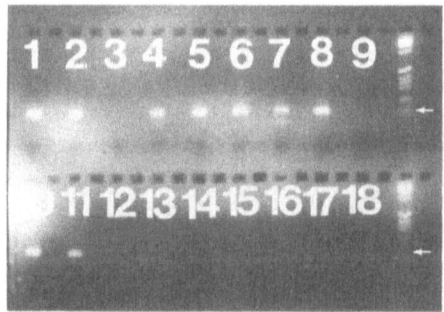

Figure 1. Analysis of clinical samples for the presence of enterovirus RNA by nested PCR. Positions 1-11, Clinical samples (See Table 1 for further details). Positions 12-18, reagent blanks. An arrow indicates the position of a 142/154 bair pair doublet present in molecular weight marker tracks.

b) Serotyping enterovirus isolates. Sequencing of PCR products provides a rapid means of comparing clinical isolates with reference strains at the molecular level. PCR may be of particular value in confirming the identity of viruses which cannot be neutralized using reference antisera (Muir et al 1993), and for characterization of novel or variant enterovirus strains and serotypes.

c) Studying enterovirus pathogenesis. We have used PCR to detect enterovirus RNA in both cryopreserved and archival formalin-fixed myocardial tissue from patients with chronic heart disease (Muir et al 1993) and this may be useful in studying the pathogenesis of persistent enterovirus infection. Some recent studies employing PCR have found no increased prevalence of enterovirus RNA in myoacrdial tissue from patients with dilated cardiomypathy (Petitjean et al 1992; Keeling et al 1993; Liljeqvist et al 1993) and polymyositis (Leff et al 1992). This is in contrast to results from studies employing slot blot or *in situ* hybridization techniques (Bowles et al 1986; 1987; Kandolf et al 1990; Yousef et al 1990). This may be due to differences in the relative sensitivity and specificity of PCR and hybridization techniques, or differences in the patient groups studies. Further studies employing both PCR and hybridization techniques of known sensitivity to study tissue from the same patients is required to resolve this discrepancy. We are currently involved in quantitation studies using infected mouse tissue to compare the relative senisitivty of PCR and in situ hybridization for the detection of enterovirus RNA in cryopreserved and formalin-fixed tissue. PCR may also be useful for investigating the role of persistent enterovirus infection in other chronic diseases including chronic fatigue syndrome (Gow et al 1991) and post-poliomyelitis muscular atrophy (Sharief et al 1991).

d) **Monitoring therapy in patients with enterovirus-related disease.** At present specific antiviral agents or immunotherapeutic treatments are not available for enterovirus infections, although this is being researched. Should they become available, a rapid procedure to detect enteroviruses would be of help in both clinical trials and routine use of specific therapy.

e) **Monitoring enteroviruses in the environment.** PCR may be useful in monitoring enteroviruses in sewage and public water supplies and utilities. This may be particularly useful in assessing the effectiveness of poliovirus eradication measures in developing countries.

REFERENCES

Bowles NE, Dubowitz V, Sewry CA and Archard LC. Dermatomyositis, polymyositis, and coxsackie-B-virus infection. Lancet 1987, i: 1004-7

Bowles NE, Richardson PJ, Olsen EGJ and Archard LC. Detection of coxsackie-B-virus-specific RNA sequences in myocardial biopsy samples from patients with myocarditis and dilated cardiomyopathy. Lancet 1986, i: 1120-3

Gow JW, Behan WMH, Clements GB, Woodall C, Riding M and Behan PO. Enteroviral RNA sequences detected by polymerase chain reaction in muscle of patients with postviral fatigue syndrome. Br Med J 1991, 302: 692-6

Kandolf R, Canu A, Klingel K, Kirschner P, Schonke H, Mertsching J, Zell R and Hofschneider PH. Molecular studies on enteroviral heart disease. In: Brinton MA, Heinz FX (eds) New aspects of positive strand RNA viruses. American Society of Microbiology, Washington, 1990, pp 340-8

Keeling PJ, Jeffery S, Caforio ALP, Taylor R, Bottazzo GF, Davies MJ and McKenna WJ. Similar prevalence of enteroviral genome of patients with idiopathic dilated cardiomyopathy and controls by the polymerase chain reaction. Brit Heart J 1992, 68: 554-9

Leff RL, Love LA, Miller FW, Greenberg SJ, Klein EA, Dalakas MC and Plotz PH. Viruses in idiopathic inflammatory myopathies: absence of candidate viral genomes in muscle. Lancet 1992, 339: 1192-5

Liljeqvist J-A, Bergstrom T, Holmstrom S, Samuelson A, Yousef, GE, Waagstein F and Jeansson S. Failure to demonstrate enterovirus aetiology in Swedish patients with dilated cardiomyopathy. J Med Virol 1993, 39: 6-10

Melnick JL. Enteroviruses: polioviruses, coxsackieviruses, echoviruses and newer enteroviruses. In: Fields BN, Knipe DM, Chanock RM, Hirsch MS, Melnick JL, Monath TP, Roizman B (eds) Virology, 2nd edition, Raven Press, New York, 1990, pp 549-605

Muir P, Kandolf R. Laboratory diagnosis of enteroviral heart disease. In: Banatvala JE (ed) Viral infections of the heart, Edward Arnold, London (1993)

Muir P, Nicholson N, Jhetam M, Neogi S and Banatvala JE. Rapid diagnosis of enterovirus infection by magnetic bead extraction and polymerase chain reaction detection of enterovirus RNA in clinical specimens. J Clin Microbiol 1993, 31: 31-38

Nicholson F, Meetoo G, Aiyar S, Banatvala JE and Muir P. Detection of enterovirus RNA in clinical samples by nested polymerase chain reaction for rapid diagnosis of enterovirus infection. J Virol Meth 1994, in press

Petitjean, J, Kopecka H, Freymuth F, Langlard JM, Scanu P, Galateau F, Bouhour JB, Ferriere M, Charbonneau P and Komajda M, Detection of enteroviruses in endomyocardial biopsy by molecular approach. J Med Virol 1992, 37: 76-82

Rotbart HA. New methods of rapid enteroviral diagnosis. Prog Med Virol 1991, 38: 96-108.

Sharief, MK, Hentges R and Ciardi M. Intrathecal immune response in patients with the post-polio syndrome. N Engl J Med 1991, 325: 749-755

Yousef GE, Isenberg DA and Mowbray JF. Detection of enterovirus specific RNA sequences in muscle biopsy specimens from patients with adult onset myositis. Ann Rheum Dis 1990, 49: 310-315

MONITORING OF POLIOVIRUS STRAINS CIRCULATING IN THE ENVIRONMENT AND THEIR INTERTYPIC AND INTRATYPIC DIFFERENTIATION BY RFLP ANALYSIS OF A PCR PRODUCT DERIVED FROM THE 5' NON-CODING REGION

B. Schweiger, B. Böhtig, S. Diedrich, J.M. López-Pila* and E. Schreier

Robert-Koch-Institut des Bundesgesundheitsamtes, *Institut für Wasser-, Boden- und Lufthygiene des Bundesgesundheitsamtes, Berlin, Germany

Efforts to eradicate poliomyelitis encompass not only an appropriate vaccination policy but steps to verify the success of eradication campaigns as well. Conventional procedures for monitoring the circulation of virulent polio strains include the isolation of enteroviruses from sewage, their differentiation as polio or non-polio strains, and their intertypic and intratypic differentiation. We have developed a procedure which considerably facilitates this task: after a strain has been isolated and identified as polio, we extract its RNA, reverse transcribe it to cDNA, carry out PCR and analyze the amplified fragment with various restriction enzymes.

The restriction pattern found allows the differentiation between the three serotypes as well as the differentiation of a given serotype as attenuated (Sabin-like) or wild (non Sabin-like). More than 60 isolates have been analyzed in this manner, the results demonstrating the usefulness of the procedure for environmental monitoring

INTRODUCTION

Some important epidemiological questions require the monitoring of pathogens in the free environment. For instance, the goal of the World Health Organization to eradicate poliomyelitis by the year 2000 includes the monitoring of sewage for the presence of virulent strains of the poliomyelitis viruses. This task, however, poses a challenge to public health laboratories because the detection with conventional procedures of very small concentrations of virulent strains in sewage requires considerable efforts and manpower.

Methods in DNA Amplification, Edited by
A. Rolfs *et al.*, Plenum Press, New York, 1994

Usually, after the isolation by cell culture of a number of enteroviruses from sewage, these are identified as polio or non-polio by seroneutralisation. The polio isolates are then differentiated in one of the three serotypes by seroneutralisation (intertypic differentiation), and further differentiated as Sabin-like or virulent (intratypic differentiation). Widely used methods for intratypic differentiation are neutralization with strain specific (McBride 1959; Wecker 1960) or cross absorbed neutralizing antisera (van Wezel and Hazendonk 1979) and phenotypic markers, such as the rct marker (ability to grow at supraoptimal temperature) (Lwoff and Lwoff 1958).

Aiming at simplifying the steps for the identification of the isolates, we have attempted to substitute the procedures involving cell cultures with a restriction fragment length polymorphism assay (RLFP), which takes advantage of the genomic differences between the strains. We have selected an appropriate fragment of the polio genome and within the cDNA of this fragment we have identified all available restriction sites in the three Sabin serotypes. We have then exposed the PCR products of the Sabin strains and of confirmed wild type isolates to the corresponding restriction enzymes and have selected those enzymes which allow intertypic and intratypic differentiation.

MATERIAL AND METHODS

Viruses. The overall majority of the polio strains used here were German isolates. The other strains were kindly supplied by Dr. Hovi, Helsinki, Finland (P3/Finland/23127/84), Dr. Lipskaya, Moscow, Russia (P1/Lugovskoi, P3/RJA-8, P3/ WL 10/2, P3/JR 28/1), Dr. Combiecsu, Bucarest, Romania (P1/France/79), and Dr. Ganbator, Ulan Bator, Mongolia (P3/0580/89). Viruses were propagated in RD cell monolayer cultures. Differentiation of polioviruses was performed using the antigenic marker test (McBride 1959), the rct marker (Lwoff and Lwoff 1958), the elution marker (Szöllösy et al 1969; Thomssen and Mayer 1965), and oligonucleotide fingerprinting (Nottay et al 1981).

RT-PCR and characterization of PCR products. The primers used corresponded to the nucleotide numbers 162-182 <5' CAA GCA CTT CTG TTT CCC CGG 3'> and 577-596 <5' TGG CTG CTT ATG GTG ACA AT 3'> of the polio 1 genome, Mahoney strain. For liberating the poliovirus genome, 2 µl of infected cell culture were heated to 65 °C for 30 min and subsequently used in the reverse transcription reaction (RT). The viral RNA was reverse transcribed with the antisense primer (bp 577-596) using 100 units of Moloney murine leukaemia virus reverse transcriptase at 37 °C for 1 hour. After initial 3 minutes denaturation at 96 °C the DNA was amplified by 40 cycles of PCR (30 sec 96 °C, 30 sec 53 °C, 1 min 72 °C) adding sense primer 2 and 1 unit of Taq DNA polymerase (Perkin-Elmer, Cetus). The reaction products were analyzed on an 1.5% agarose gel containing ethidium bromide. For restriction enzyme analysis 5 µl of the PCR product were digested with the appropriate enzyme in 15 µl volumes. Incubation was as recommended by the supplier (Boehringer, Mannheim). The digests were analyzed on a 2% agarose gel. For Southern blot and nucleic acid hybridization uncleaved and enzymatic digested PCR products were transferred to a nylon membrane and hybridized against a Digoxigenin (DIG) labeled probe, which had been prepared by PCR amplification of Sabin 1 cDNA between the basepairs 186 and 468, followed by labeling the product with DIG-dUTP. After hybridization at 42 °C for 16 h DIG chemiluminescent detection was performed according to the instructions of the supplier (Boehringer, Mann-heim).

Table 1. Restriction enzymes used for the differentiation of vaccine and wild polioviruses

Strain	Enzyme	Fragment size (bp)
Sabin 1	NcoI	207, 227
	AvaI	118, 314
	AvaII	190, 294
	BamHI	59, 375
	HaeIII	56, 80, 89, 209
Sabin 2	SmaI	140, 295
	ClaI	58, 377
	HaeIII	80, 146, 209
	HpaII	107, 121, 149
	BsiYI	78, 82, 94, 138
Sabin 3	NcoI	124, 314
	HpaII	121, 148, 151
	MluI	115, 323
	PvuII	110, 328
	HaeIII	54, 70, 129, 145

RESULTS AND DISCUSSION

PCR and restriction enzymes selected for the RFLP-assay. On the basis of sequence data of the 5'-NCR of the poliovirus genome, restriction enzymes were selected to allow an unambiguous identification of the three Sabin serotypes and to distinguish between Sabin-like and wild-type strains. The enzymes used for differentiation of vaccine and wild type viruses and the corresponding restriction fragments of the Sabin strains are summarized in Table 1. In this study more than 60 virus isolates including prototype strains have been investigated and, so far, successfully intratypically differentiated. In each case a 435 base pair amplification product was digested and the restriction fragments were characterized on agarose gels (Fig. 1).

RFLP-analysis of poliovirus type 1: Ten Sabin 1-like isolates, five isolates designated as "unidentified" and six wild strains were included in the study. The results demonstrate that the restriction sites for NcoI, AvaII, and BamHI are highly conserved in the Sabin-like isolates (Table 2). After digestionb with HaeIII Sabin 1-typical migration patterns were yielded for 90% of the vaccine-related isolates. Besides this characteristic Sabin 1 restriction profile, an additional fragment was observed for two isolates (Table 2). Cleavage ba AvaI showed that the restriction site for this enzyme is present in half of the Sabin 1-like samples (Table 2). Therefore, AvaI is only for limited use and gives, if cleavage is evident, further information.With exception of AvaI, no different migration patterns were found for Mahoney and Sabin 1. In general, no fragments typical for vaccine strains wre obtained by cleaving the other wild strains tested with these five enyzmes (Table 2). The strains belonging to the group designated as "unidentified isolates" had already been analyzed by conventional methods and had shown nonvaccine-like charactersitics. RFLP-analysis revealed vaccine-liked patterns for three or more enyzmes indicating that these isolates were in fact vaccine-dertived (Table 2).

Table 2. Restriction enzyme analysis of type 1 polioviruses. [a]Restriction patterns are represented as follows: +: Typical pattern for Sabin 1, -: no cleavage of other pattern than Sabin 1, +/-: typical pattern for Sabin 1 plus another unspecific fragment. [b]Characterization by virological methods. [c]Nonvaccine-like behavior using virological methods. [d]Additionally characterized by oligonucleotide fingerprinting. [e]Environmental samples (water, sewage).

Strain	Enzyme				
	NcoI	AvaII	BamHI	HaeIII	AvaI
Refernce viruses					
Sabin I (LSc 2ab)	+	+	+	+	+
Mahoney/USA42	+	+	+	+	+
Vaccine-related isolates[b]					
9510/63	+	+	-	+/-	+
9181/63	+	+	+	-	-
0129/92	+	+	+	+	+
0043/92	+	+	+	+/-	+
0184/91[e]	+	+	+	+	+
0186/91[e]	+	+	+	+	+
0128/91[e]	+	+	+	+	-
0217/92[e]	+	+	+	+	-
0219/92[e]	+	+	+	+	+
0220/92[e]	+	+	+	+	-
Unidentified isolates[c]					
0097/91	+	+	+	+	-
0106/91	+	+	+	+/-	-
0107/91	+	+	+	-	-
0110/91	+	+	+	-	-
0101/91	+	+	-	+	-
Wild isolates[b]					
Lugovskoi[d]	-	-	-	-	-
France/79[d]	-	+	-	-	-
4784/61	-	-	-	-	-
1040/49	-	-	-	-	-
2594/59	-	-	-	-	-

RFLP-analysis of poliovirus type 2. Twelve Sabin 2-like isolates and eight wild type strains were selected for restriction analysis. The enzymes SmaI, ClaI, HpaII, BsiYI, and HaeIII generated for all of the twelve isolates the same patterns as they did for Sabin 2 (Table 2). The wild strains used in this study lacked a SmaI and ClaI cleavage site. All wild strains tested were cleaved by HpaII and BsiYI, but the migration patterns were quite different compared to Sabin 2. With HaeIII we obtained for Sabin 2 and the majority of the wild strains the same results. For this reason HaeIII is not suitable. However, the enzymes SmaI, ClaI, HpaII and BsiYI differentiated very reliably between the Sabin-like and the wild type viruses assayed (Table 3).

Table 3. Restriction enzyme analysis of type 2 polioviruses[a]. aRestriction patterns are represented as follows: +: typical pattern for Sabin 2, -: no cleavage or othern pattern than Sabin 2. [b]Characterization by virological methods. [c]Isolated before introduction of Sabins'vaccine

Strain	Enzyme				
	SmaI	ClaI	HpaII	BsiYI	HaeIII
Refernce viruses					
Sabin 2 (P712 Ch.2ab)	+	+	+	+	+
MEF-1/EGY42	-	-	-	-	+
Vaccine-related isolates[b]					
0917/59	+	+	+	+	+
093/77	+	+	+	+	+
0206/91	+	+	+	+	+
0137/91	+	+	+	+	+
0207/91	+	+	+	+	+
038/91	+	+	+	+	+
0117/91	+	+	+	+	+
0232/91	+	+	+	+	+
0293/91	+	+	+	+	+
04/91	+	+	+	+	+
063/92	+	+	+	+	+
0266/92	+	+	+	+	+
Wild isolates[c]					
318/58	-	-	-	-	-
486/58	-	-	-	-	+
522/58	-	-	-	-	+
922/59	-	-	-	-	+
1090/59	-	-	-	-	-
2506/59	-	-	-	-	+
2557/59	-	-	-	-	-

RFLP-analysis of poliovirus type 3. Ten Sabin 3-like isolates and eight wild type strains were analyzed. The PCR prodcuts were digested with five different enyzmes. All of the Sabin 3-like samples cleaved by NcoI, HpaII, and MluI showed restriction patterns expected for the vaccine virus (Table 4). With PvuUU and HaeIII Sabin 3-typical cleaving prodcuts were also evident for these isolates, but in some cases an additional fragment was observed (Table 4).The restriction analysis of the wild type 3 strains assayed showed NcoI, HpaII, and HaeIII generated patterns quite different from those of the vaccine strains (Table 4).The digestion with MluI yielded, one strain expected, also no Sabin 3-typical fragments. PvuII generated for some of the wild strains the same migration patterns as for Sabin 3. However, because four out of five enyzmes produced patterns not characteristic of the vaccine virus, differentiation between vaccine-related isolates and wild type virus was always feasible.

Table 4. Restriction enzyme analysis oftype 3 poliovirusa. [a]Restriction patterns are represented as follows: +: typical pattern for Sabin 3, -: no cleavage or othern pattern than Sabin 3, +/-: typical pattern for Sabin 3 plus another unspecific fragment. [b]Characterization by virological methods. [c]Characterization by virological methods and oligonucleotide fingerprinting. [d]Environmental samples (sewage).

Strain	Enzyme				
	NcoI	HpaII	MluI	PvuII	HaeIII
Refernce viruses					
Sabin 3 (Leon 12a,b)	+	+	+	+	+
Saukett/USA52	-	-	+	-	-
Vaccine-related isolates[b]					
1011/68	+	+	+	+	+/-
0330/91	+	+	+	+	+
0114/91	+	+	+	+	+
0111/91	+	+	+	+	+
05/92	+	+	+	+/-	-
0141/91	+	+	+	+	+
0214/91[d]	+	+	+	+	+
0215/91[d]	+	+	+	+/-	+/-
0216/92[d]	+	+	+	+	+
0218/92[d]	+	+	+	+	+
Wild isolates[c]					
Finland/84	-	-	-	-	-
RJA-8	-	-	-	+	-
WL 10/2	-	-	-	+	-
JR 28/1	-	-	-	+	-
0580/89	-	-	-	-	-
042/90	-	-	-	-	-
045/92	-	-	-	-	-

DISCUSSION

For differentiation of vaccine-derived and wild strains many serological and physicochemical marker tests have been developed, but none of these known markers is uniquely associated with vaccine viruses and their derivates (Crainic et al 1983; Nakano et al 1978). Because a rapid and specific assay for intratypic differentiation would be an useful supplement to clinical and virological diagnostic procedures, we describe a method to differentiate polioviruses using amplification of viral cDNA followed by selective restriction enzyme digestion.

Recently, poliovirus genotypes have been differentiated by PCR followed by a restriction fragment length polymorphism (RFLP) assay similar to the one described here. The genomic segment analyzed was located in a highly variable region corresponding to the VP1 subunit (Balanant et al 1991). The three major capsid proteins VP1, VP2, VP3, all exhibit some kind of surface exposure and contain most of the antigenic sites of the virus that can be modified

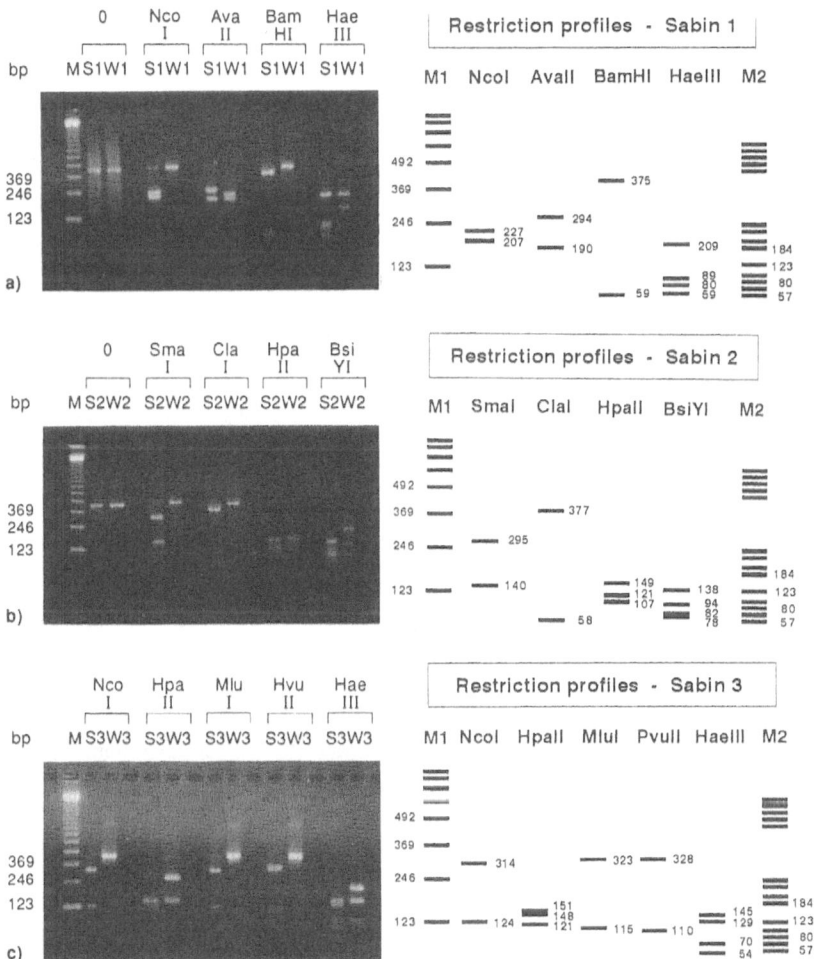

Figure 1. Agarose gel analysis (left) and schematic representation (right) of amplified cDNA of poliovirus strains and of the corresponding restriction fragments. a) Type 1 strains. S1: Sabin 1; W1: wild type strain 2594/59. b) Type 2 strains. S2: Sabin 2; W2:MFE-1. c) Type 3 strains. S3: Sabin 3; W3: Finland/84. Abbreviations: O undigested PCR product, M marker.

rapidly under immune selection (reviewed by Hogle and Filman 1989). Additionally, all mutated residues that are less accessible to antibody interaction are located close to fully exposed sites of the virus (Page et al 1988). Choosing restriction sites for diagnosis in this variable region might therefore generate misleading results due to mutations affecting the restriction sites. This encouraged us to investigate a region of the virus which is less exposed to the selective pressure of neutralizing antibodies.

As the genomic region to be amplified and analyzed we choosed the 5'-NCR, which has been shown to be conserved among all enteroviruses (Chang et al 1989; Hughes et al 1986,1989; Iizuka et al 1987; Jenkins et al 1987; Kitamura et al 1981; La Monica et al 1986; Lindberg et al 1987; Nomoto et al 1982; Rancaniello and Baltimore 1981; Ryan et al

1990; Stanway et al 1984; Toyoda et al 1984). We decided on this region to ensure PCR amplification for the wild-type strains as well, with one pair of universal primers. Although high mutation frequencies have been shown for polioviruses (De la Torre et al 1990; Ward et al 1988; Ward and Flanagan 1992), we expected that the vaccine strains would not have changed significantly in the 5'-NCR. The results obtained with a set of various restriction enzymes have so far confirmed highly conserved cleavage sites for vaccine-derived isolates. Moreover, sequencing of the 5'-NCR of ten other Sabin 3-related strains in our laboratory (Driesel et al submitted for publication) also corroborated the restriction sites for the enzymes reported here, excepted for four strains showing mutations at the restriction site for HaeIII. Besides the HaeIII site, other mutations in the 5'-NCR were found ranging from two to ten nucleotides.

For differentiation between vaccine-like and wild type virus, genotype-specific amplification by PCR (Yang et al 1991, 1992) or the use of genotype-specific hybridization probes (da Silva et al 1991) have been described. However, for routine monitoring purposes these procedures might yield inconclusive results when the isolates contain heterogeneous populations of wild and vaccine strains. Moreover, only those wild strains are detectable for which sequence information is available for preparing the specific primer or probes. With the assay described here, if a particular isolates consists of a mixture of a virulent and a Sabin-like strain, it would be apparant from the restriction profile and the electrophoresis pattern, respectively.

REFERENCES

Balanant J, Guillot S, Candrea A, Delepeyroux F, and Crainic R. The natural genomic variability of poliovirus analyzed by a restriction fragment length polymorphism assay. Virology 1991, 184:645-654.

Chang KH, Auvinen P, Hyypiä T, and Stanway G. The nucleotide sequence of coxsackievirus A9; implications for receptor binding and enterovirus classification. J Gen Virol 1989, 70:3269-3280.

Crainic R., Couillin P, Blondel B, Cabau N, Bouè A, and Horodniceanu F. Natural variation of poliovirus neutralization epitopes. Infec. Immun 1983, 41:1217-1225.

da Silva EE, Pallansch MA, Holloway BP, Oliveira MJC., Schatzmayr HG, and Kew OM. Oligonucleotide probes for the specific detection of the wild poliovirus types 1 and 3 endemic to Brazil. Intervirology 1991, 32:149-159.

De La Torre JC, Wimmer E, and Holland JJ. Very high frequency of reversion to guanidine resistance in clonal pools of guanidine-dependent type 1 poliovirus. J Virol 1990, 64:664-671.

Driesel G, Künkel U, Diedrich S, Böthig B, and Schreier E. Submitted for publication.

Hogle JM, and Filman DJ. Poliovirus: three-dimensional structure of a viral antigen. Adv. Vet. Sci. Comp. Med. 1989, 33:65-91.

Hughes PJ, Evans DMA, Minor PD, Schild GC, Almond JW, and Stanway G. The nucleotide sequence of a type 3 poliovirus isolated during a recent outbreak of poliomyelitis in Finland. J Gen Virol 1986, 67:2093-2102.

Hughes PJ, North C, Minor PD, and Stanway G. The complete nucleotide sequence of coxsackievirus A21. J Gen Virol 1989, 70:2943-2952.

Iizuka N, Kuge S, and Nomoto A. Complete nucleotide sequence of the genome of coxsackievirus B1. Virology 1987, 156:64-73.

Jenkins O, Booth JD, Minor PD, and Almond JW. The complete nucleotide sequence of coxsackievirus B4 and its comparison to other members of the picornaviridae. J Gen Virol 1987, 68:1835-1848.

Kitamura N, Semler BL, Rothberg MG, Larsen GR, Adler CJ, Dorner AH, Emini AE, Hanecak R, Lee JJ, Van Der Werf S, Anderson CW, and Wimmer E. Primary structure, gene organization and polypeptide expression of poliovirus RNA. Nature 1981, 291:547-553.

LaMonica N, Meriam C, and Racaniello VC. Mapping of sequences required for mouse neurovirulence of poliovirus type 2 Lansing. J Virol 1986, 57:515-525.

Lindberg AM, Stalhandske POK, and Pettersson U. Genome of coxsackievirus B3. Virology 1987, 156:50-63.

Lwoff A, and Lwoff M. L'inhibition du développement du virus poliomyélitique á 39° et le probléme du role de l'hyperthermie dans l'evolution des infections virales. C r hebd Séanc Acad Sci, Paris 1958. 1958, 246:190-192.

McBride WD. Antigenic analysis of polioviruses by kinetic studies of serum neutralization. Virology 1959.7:45-58.

Nakano, JH, Hatch MH, Thieme ML, and Nottay B. Parameters for differentiating vaccine-derived and wild poliovirus strains. Prog Med Virol 1978, 24:178-206.

Nomoto A. Omata T, Toyoda H, Kuge S, Horie H, Kataoka Y, Genba Y, Nakano Y, and Imura N. Complete nucleotide sequence of the attenuated poliovirus Sabin 1 strain genome. Proc Natl Acad Sci USA 1982, 79:5793-5797.

Nottay BK, Kew OM, Hatch MH, Heyward JT, and Obijeski JF. Molecular variation of type 1 vaccine-related and wild polioviruses during replication in humans. Virology 1981, 108:405-423.

Page GS, Mosser AG, Hogle JM, Filman DJ, Rueckert RR, and Chow M. Three-dimensional structure of poliovirus serotype 1 neutralizing determinants. J Virol 1988, 62:1781-1794.

Racaniello VR, and Baltimore D. Molecular cloning of poliovirus cDNA and determination of the complete nucleotide sequence of the viral genome. Proc Natl Acad Sci USA 1981, 78:4887-4891.

Ryan MD, Jenkins O, Hughes PJ, Brown A, Knowles NJ, Booth D., Minor PD, and Almond JW. The complete nucleotide sequence of enterovirus type 70: relationships with other members of the picornaviridae. J Gen Virol 1990, 71:2291-2299.

Stanway, G, Hughes PJ, Mountford RC, Reeve P, Minor PD, Schild GC, and Almond JW. Comparison of the complete nucleotide sequences of the genomes of the neurovirulent poliovirus P3/Leon/37 and its attenuated Sabin vaccine derivative P3/Leon 12a₁b. Proc Natl Acad Sci USA 1984, 81:1539-1543.

Szöllösy E, Hils J, Böthig BB. Über die Verwendbarkeit von ^{32}P-markierten Polioviren, gezüchtet in verschiedenen Zellkulturen, für chromatographische Untersuchungen. Dtsch Gesundh wes 1969, 49:2333-2337.

Thomssen R, and Mayer M. Different pattern of elution of poliovirus strains from DEAE cellulose and aluminiumhydroxide gel. Arch gesamte Virusforsch 1965, 15:735.

Toyoda H, Kohara M, Kataoka Y, Suganuma T, Omata T, Imura N, and Nomoto A. Complete nucleotide sequences of all three poliovirus serotype genomes. Implication for genetic relationship, gene function and antigenic determinants. J Mol Biol 1984, 174:561-585.

van Wezel AL., and Hazendonk AG. Intratypic serodifferentiation of poliomyelitis virus strains by strain-specific antisera. Intervirology 1979, 11:2-8.

Ward CD, Stokes MAM and Flanegan JB. Direct measurement of the poliovirus RNA polymerase error frequency in vitro. J Virol 1988, 62:558-562.

Ward CD, and Flanegan JB. Determination of the poliovirus RNA polymerase error frequency at eight sites in the viral genome. J Virol 1992, 66:3784-3793.

Yang CF, De L, Holloway BP, Pallansch MA, and Kew. OM. Detection and identification of vaccine-related polioviruses by the polymerase chain reaction. Virus Res 1991, 20:159-179.

Yang CF, De L, Yang SJ, Gómez JR, Cruz JR, Holloway BP, Pallansch MA, and Kew OM. Genotype-specific in vitro amplification of sequences of the wild type 3 polioviruses from Mexico and Guatemala. Virus Res 1992, 24:277-296.

DETECTION OF ADENOVIRUSES IN BLOOD BY POLYMERASE CHAIN REACTION

M. Dahme, W. Seidel, G. Flunker, A. Peters, S. Wiersbitzky and H. Reddemann

Institut für Medizinische Mikrobiologie, Abtl. für Virologie und Klinik und Poliklinik für Kindermedizin, Ernst-Moritz-Arndt, Universität Greifswald, Germany

INTRODUCTION

Adenoviruses have been recovered from virtually every organ system of man and have been associated with many clinical syndromes. There are at present 47 accepted serotypes of human adenoviruses. Many adenovirus infections are subclinical and result in antibody formation that is probably protective against an exogenous reintroduction of the same adenovirus serotype. The most common types are 1 - 8, 11, 21, 35, 37 and 40 (Hierholzer 1991, Schmitz et al 1983). Several groups of investigators, who have found low levels of adenovirus DNA persisting in peripheral blood lymphocytes, have suggested that lymphocytes have a role in the maintenance of persistence. It is known that Ad 1, Ad 2, Ad 5 and Ad6 (subgenus C) can persist in tonsils and adenoids for several years (Andiman and Miller 1982, Horvath et al 1986, Lavery et al 1987). Increasing incidence of adenovirus disease has been reported in primary immundeficiens and in organ transplant recipients (Hierholzer 1992, Flomenberg et al 1994). The results of other studies suggest that immunocompromised patients may be no more commonly infected with adenoviruses than normal hosts, but the outcome may be more serious and the illness may more likely be fatal (Cames et al 1992, Anders et al 1991, Teague et al 1991). Ad 1, Ad 2 and Ad 6 of subgenus C occurred among immunocompromised children, whereas Ad 4 and Ad 5 caused severe infection in adults (Hierholzer 1991). Persistent adenoviruses might explain the occasional clinical illness in immunocompromised hosts who could experience reaction of a latent infection at the time of immunosuppression.

The current methods for detecting persistent adenovirus infection, such as culture, antigen or antibody detection, are usually inadequate. Therefore we used the polymerase chain reaction to investigate the lymphocytes and whole blood of immunocompromised children for persistent adenovirus DNA.

Methods in DNA Amplification, Edited by
A. Rolfs *et al.,* Plenum Press, New York, 1994

MATERIALS AND METHODS

Peripheral blood specimens were drawn from 33 immunocompromised patients (aged 1 to 16 years) and submitted to the Regional Diagnostic Laboratory at the University Hospital of Greifswald between January and December 1992.

Measured aliquots (25 µl) of each anticoagulated whole-blood sample were spotted onto a precut filter paper. Dried blood spots were stored at room temperature up to 9 months until required for analysis.

The DNA was eluted from the dried blood spots by incubating each filter paper in 0.5 ml of digestion buffer (10mM Tris pH 8.0, 10mM EDTA, 50mM NaCl, 2% SDS, 10µg Proteinase K [Boehringer Mannheim] per ml) for 2 hours at 56 °C. The sample was extracted twice with an equal volume of phenol-chloroform-isoamyl alcohol. The DNA was precipitated for 30 min at -70 °C in 0,25M ammonium acetate, 10µg per ml of the carrier Escherichia coli t RNA (Boehringer Mannheim) and 2 volumes of ice-cold ethanol. The air-dried DNA pellet was reconstituted in 100µl of sterile H_2O after centrifugation and washed with 70% ethanol and an aliquot (2µl in a 50µl reaction mixture) was amplified in the PCR.

The mononuclear cell fraction was separated from the remaining portion of each sample by centrifugation on Ficoll-Paque (Pharmacia). The lymphocytes were resuspended in 0,5 ml lysis buffer (0.3 M sucrose, 10mM Tris pH 7.5, 5mM $MgCl_2$, 1% Triton X-100) and the nuclei were pelleted. The nuclei were resuspended in 100µl PCR buffer (50mM KCl, 10mM TRIS pH 8.3, 2.5mM $MgCl_2$ 0.1 mg/ml gelatine, 0.45 % NP 40, 0.45% Tween 20) and treated with 100µg/ml Proteinase K for 4 hours at 56 °C. An aliquot (1µl in a 50µl reaction mixture) of the sample was amplified in the PCR after inactivation of Proteinase K for 20 min at 95 °C.

PCR

Two sets of primers including general primers specific for the hexon-coding region (H1-H4) of all six subgenera and specific primers for the fiber-coding region (P1-P4) for the subgenera C were applied to a nested PCR (Figure 1).

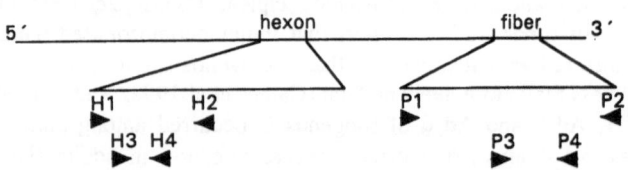

Figure 1.   Location and sequences of the four sets of oligonucleotide primers. Position, sequence and fragment length of the amplicons created in the study: H_1 5'GCCGCAGTGGTCTTACATGCACATC 3' (18858-18882), H_2: 3'TGAAACTGTAGGCGCCGCACGAC 5' (19136-19158, H1/H2: 300bp) H_3 5'CCACC GAGACGTACTTCAGCC 3' (18938-18958), H_4 3'CGACACCCACTGTTGGCACAC 5'(19096-19116, H3/H4: 178bp); P_1: 5'ATGAAACGCGCCAGACC 3' (31030-31046), P_2: 3'AGCGAGTAGTCACTATAATT 5' (33065-33084, P1/P2: 2,054bp); P_3: 5'ATGTAACGGGTCCTTATTT 3' (32761-32779), P_4: 3'TGTG GACAACCCCAGACCCATCTCCTAACTG 5' (32231-32261, P3/P4: 548bp)

The general primer pairs H1, H2 (Allard et al., 1990) and H3, H4 were used in a PCR for 30 cycles at 95 °C for 10 sec, 55 °C for 30 sec and 72 °C for 30 sec. The DNA was amplified in a 50µl reaction mixture containing 250µM of each dNTP, 20pM of each primer, Taq-DNA-polymerase buffer (Amersham) and 0.5U Taq-DNA-polymerase (Amersham). 1µl of the first PCR product (using the H1, H2 primers) was applied to the nested PCR (using the H3, H4 primers).

The specific external primer set (P1, P2) was used at 95 °C for 10 sec, 57 °C for 90 sec, 72°C for 90 sec and the specific internal primer set (P3, P4) at 95 °C for 10 sec, 53 °C for 90 sec, 72 °C for 90 sec, both for 30 cycles and under the same conditions as for the hexon primers. 1µl of the first PCR product (using the P1, P2 primers) was applied to the nested PCR (using the P3, P4 primers).

The PCR products were analysed in a 2% agarose gel (after ethidium bromide staining), transferred to a nylon membrane (Amersham) and detected with a ^{32}P-labelled probe.

RESULTS

We studied the sensitivity of our assays to detect adenovirus DNA in different human materials. The sensitivity of the different sets of primers differed concerning the absolute numbers of molecules that were detected in the PCR. When pure adenovirus type 2 DNA was amplified by the specific fiber primers (P1, P2) it was possible to detect 1.5×10^5 adenovirus DNA molecules by PCR, whereas the sensitivity of the assay with the general hexon primers (H1, H2) increased 100 fold to 1.5×10^3 DNA molecules. A two-step amplification using the internal primers lowered the detection limit to 15 molecules for the general hexon primers (H3, H4) and to 1.5×10^3 for the specific fiber primers (P3, P4) (Figure 2).

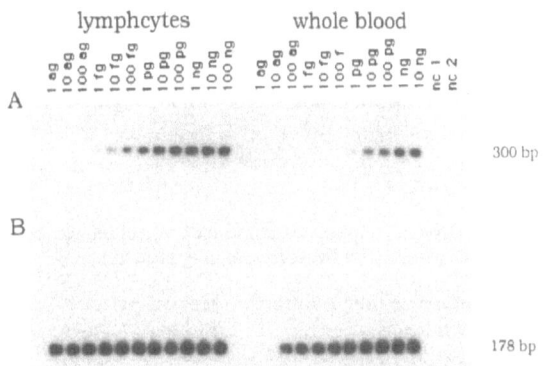

Figure 2. Sensitivity of PCR for detecting adenovirus DNA using pure adenovirus type 2 DNA: comparison between two extraction methods. A: Adenovirus type 2 DNA was amplified with the external hexon primers (H1, H2) instead of extracted DNA (same treatment as for lymphocytes or whole blood); nc 1: negative control nested PCR, nc 2: negative control second PCR; B: Aliquots of the first amplification were put in the second PCR with the internal hexon primers (H3, H4).

When we compared the contribution of the two extraction methods to sensitivity of our assay, we found that the lymphohocyte DNA extraction method resulted in a 10-fold higher sensitivity by PCR than the whole blood DNA extraction method tested by pure adenovirus type 2 DNA (Figure 2). An additional confirming hybridization step with the ^{32}P-labelled probe (548 bp or 178 bp amplification product labelled by the Megaprime DNA labelling system, Amersham) was not necessary for the sensitivity but for the specificity. All amplification products which were seen in agarose gel after ethidium bromide staining hybridized with the labelled probe (Figure 3).

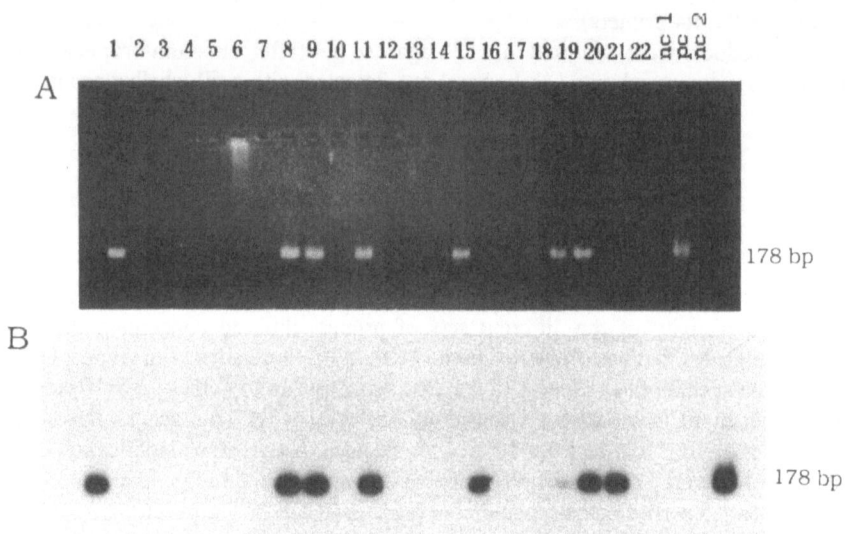

Figure 3. Detection of adenovirus DNA by a two-step PCR using the general hexon primers. A: Electrophoretic analysis of PCR products with 2% agarose gel after ethidium bromide staining. Lane 1-22 patients, nc 1: negative control nested PCR, nc 2: negative control second PCR, pc: positive control (100 pg of pure adenovirus type 2 DNA). B: Autoradiography after Southern blot transfer and hybridization with ^{32}P-labelled 178 bp fragment

Table 1. Detection of adenovirus DNA in lymphocytes and whole blood of immunocompromised patients using the two primer sets (*only positive results with the corresponding primers).

	Hexon primer*	Fiber primer*	Hexon primer Fiber primer	Negative PCR	No. of patients
Lymphocyte DNA fraction	9 27%	2 6%	3 9%	19 58%	33
Whole blood DNA fraction	9 14%	1 2%	6 9%	48 75%	64

64 specimens from 33 patients were analysed for adenovirus DNA using PCR to amplify the extracted whole-blood DNA, and 33 specimens from 22 patients to amplify the DNA from lymphocytes. 48 whole blood specimens (75%) were negative and 16 (25%) were positive for adenovirus DNA. 19 specimens (58%) were negative and 14 (42%) were positive for adenovirus DNA detected in the DNA extracted from the lymphocytes (Table 1).

None of the specimens was positive for adenovirus DNA by a one-step amplification using the hexon primers (H1, H2) or the fiber primers (P1,P2).

To investigate the possible persistence of adenovirus infection we examined a few patients for a longer period. We obtained positive results in a total of 5 patients: 2 months (3 cases), 3 months (1 case) and 5 months (1 case).

DISCUSSION

The comparison of the results indicates that both methods are able to detect adenovirus DNA in human peripheral blood as well as in human lymphocytes.

The results show likewise that more patients were positive with the general hexon primers than with the specific fiber primers. That results from the characteristics of the primers.

On the other hand, more specimens were positive for the lymphocyte DNA fraction than for the whole blood fraction. If the lymphocytes are the reservoir for the adenoviruses, then we could explain this fact by the different amounts of lymphocytes in the DNA extraction (1×10^7 lymphocytes in 2 ml and 1.25×10^5 lymphocytes in 25µl of peripheral human blood).

Positive results were obtained only by a two step PCR. That would suggest a low level of infected cells and could be an indication of a persistent adenovirus infection (Lavery et al 1987, Snejdarová et al 1975). This assumption is supported by the results from the cases that specimens from one person are positive for a period of a few months.

Whether a one-step PCR can distinguish between an acute and a persistent infection still needs to be investigated.

ACKNOWLEDGEMENT

This study was supported by a grant from the Deutsche Forschungsgemeinschaft.

REFERENCES

Allard A, Girones R, Juto P, Wadell G . Polymerase chain reaction for detection of adenoviruses in stool samples. J Clin Microbiol 1990, 28: 2659-2667

Anders KH, Park C-S, Cornford ME, Vinters H. Adenovirus encephalitis and widespread ependymitis in a child with AIDS. Ped Neurosurg 1991, 16: 316-320

Andiman WA, Miller G (1982) Persistent infection with adenovirus types 5 and 6 in lymphoid cells from humans and woolly monkeys. J Infect Dis 1982, 145: 83-88

Cames B, Rahier J, Burtomboy G, de Ville de Goyet J, Reding R, Lamy M, Otte JB, Sokal EM. Acute adenovirus hepatitis in liver transplant recipients. J Pediatrics 1992, 120: 33-37

Flomenberg P, Babbit J, Drobyski WR, Ash RC, Carrigan DR, Sedmak GV, Mc Aanliffe T, Camitta B, Horrowitz MM, Bunin N, Casper JT. Increasing incidence of adenovirus disease in bone marrow transplant recipients. J Infect DIs 1994, 169: 775-781

Hierholzer JC. Adenoviruses. in: Balows A, Hausler WJ jr, Herrmann KL, Isenberg HD, Shadomy HJ (eds) Manual of clinical microbiology (5th edition). American Society for Microbiology: 1991: 896-903

Hierholzer JC. Adenoviruses in the immunocompromised host. Clin Microbiol Rev 1992, 5: 262-274

Horvath J, Palkonyay L, Weber J. Group C adenovirus DNA sequences in human lymphoid cells. J Virol 1986, 59: 189-192

Lavery D, Fu SM, Lufkin T, Chen-kiang S. Productive infection of cultured human lymphoid cells by adenovirus. J Virol 1987, 61: 1466-1472

Schmitz H, Wigand R, Heinrich W. Worldwide epidemiology of human adenovirus infections. Am J Epidem 1983, 117: 455-466

Snejdarová V, VonkaV, Kutinová L, Rezácová D, Chládek V. The nature of adenovirus persistence in human adenoid vegetations. Arch. Virol. 1975, 48: 347-357

Teague MW, Glick AD, Fogo AB. Adenovirus infection of the kidney: mass formation in a patient with Hodgkin's disease. Am J Kid Dis 1991, 18: 499-502

DETECTION OF THE HPV 16 E2 GENE IN NONKERATINIZING SQUAMOUS CERVICAL CARCINOMAS BY PCR

L. Riethdorf[1], S. Riethdorf[2], Th. Löning[3], H.E. Stegner[3] and G. Lorenz[1]

[1]Institute of Pathology, [2]Institute of Microbiology, University of Greifswald; [3]Department of Gynaecological Pathology and Electron Microscopy, Clinics of Obstetrics and Gynaecology, University of Hamburg; Germany

From 35 nonkeratinizing squamous cervical carcinomas 22 HPV16 (human papillomavirus type 16) positive cases were detected by PCR followed by Southern blot hybridization with specific probes using paraffin embedded archival specimens. A 450 base pair fragment of the L1 gene was amplified with consensus primers. Furthermore the regions encoding the N-terminal domain for transcriptional activation (TAD) and the C-terminal DNA binding domain (DBD) of the E2 protein as well as the connecting region ("hinge") were amplified with corresponding primers. The protein E2 acts as a transcriptional modulator of viral promoters and therefore plays an important role in HPV-associated carcinogenesis of human cervical carcinoma. In 8 of the 22 HPV16-positive cases the whole E2 gene, in 9 cases only the DBD or "hinge" region of E2 and in 5 cases neither of them nor TAD could be amplified. We suppose that in 14 cases an integration of the viral DNA into the host cell genome and a deletion of E2 has occured.

INTRODUCTION

Cancer of uterine cervix is one of the most common malignant tumors in women. Infection with specific types of human papillomaviruses (HPV) has been considered to be the major etiological factor in the development of this disease. Up to 95 % of cervical cancers contain complete genomes or fragments of certain types of HPV. The most prevalent type is HPV16, while the frequency of HPV18 and other types is very low. HPV18 is assumed to be associated with the small cell type of cervical cancer.

Since it is generally believed that integration of viral DNA into the host cell genome may be an essential prerequisite for malignant progression, there have been several studies analysing the physical state of HPV. This state may serve as a prognostic indicator for the preneoplastic

Methods in DNA Amplification, Edited by
A. Rolfs *et al.*, Plenum Press, New York, 1994

lesions that are likely to progress to cancer. Furthermore it may offer clues for understanding the mechanisms involved in the process of cervical carcinogenesis.

In this study pairs of oligonucleotide primers specific for the E2 gene were used for the detection of the physical state of HPV 16 DNA.

MATERIALS AND METHODS

Paraffin embedded archival biopsy specimans from 35 squamous cervical carcinomas of nonkeratinizing large cell and small cell type were analysed. Pretreatment of the paraffin embedded tissues as described by Kiene et al (1992) was followed by PCR. A 450 base pair fragment of the L1 gene (Manos et al 1989) was amplified with a consensus primer pair (MY09, MY11, Perkin Elmer Cetus Corp.) for the HPV types 6, 11, 16, 18 and 33. Plasmid DNA containing the whole HPV 16 genome (de Villiers, Heidelberg) was used as positive control. After staining the gels with ethidium bromide, Southern blot hybridization was carried out. Digoxigenin-dUTP labelled consensus probes (MY1019, MY18) for HPV 6, 11, 16, 18, 33 and specific

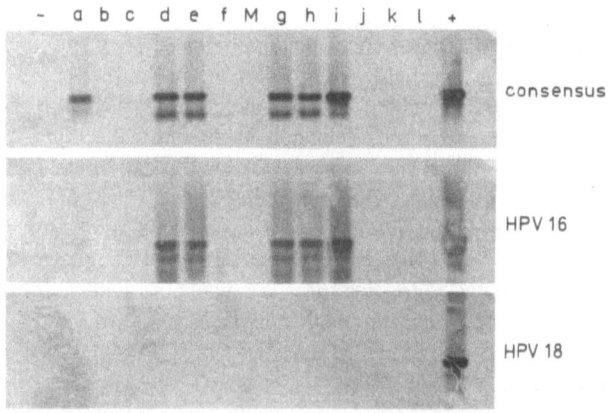

Figure 1. Southern blot hybridization with consensus probes for several HPV types (6,11,16,18,33) and specific probes for HPV 16 and 18. Hybrids were detected by the nucleic acid detection kit (Boehringer Mannheim; FRG). The cases d, e, g, h and i are HPV 16 positive. The case a shows an infection with HPV 6, 11 or 33 and is HPV 16 and 18 negative. +: positive control, -: negative control, M: unlabeled molecular weight marker

probes (MY14, WD74) for HPV16 and 18 were employed (Figure 1). Hybrids were detected by the nucleic acid detection kit (Boehringer Mannheim; FRG).

Furthermore the regions encoding the N-terminal domain for transcriptional activation (TAD) and the C-terminal DNA binding domain (DBD) of the HPV16 E2 protein as well as the connecting region ("hinge") were amplified with the corresponding primer pairs (Figure 2).

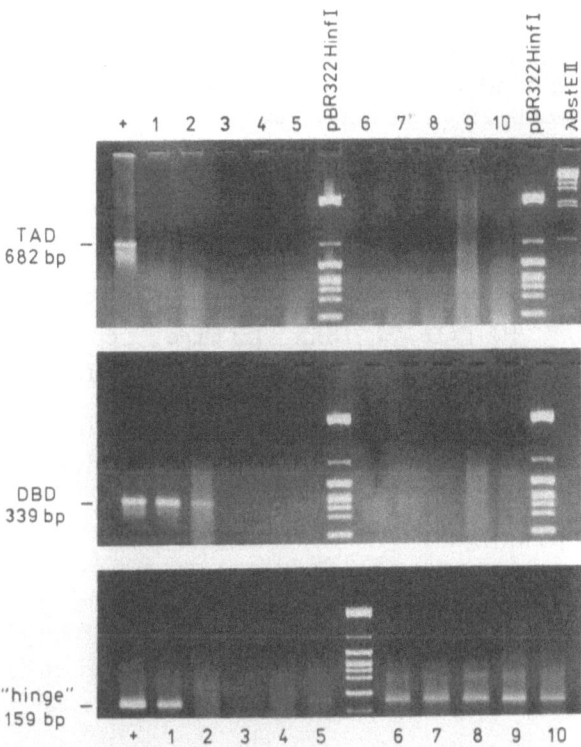

Figure 2. Amplification of the TAD-, DBD- and "hinge"-regions in several cases of cervical carcinomas by PCR. The PCR-products were analysed in 2 % agarose gels after staining with ethidium bromide. In case 1 the whole E2 gene could be amplified whereas the other cases show amplification of only parts of the E2 gene. Case 3 and 4 were negative for E2 (see also table).

RESULTS AND CONCLUSIONS

In the present study of 35 nonkeratinizing squamous cervical carcinomas 25 were found to be HPV positive by PCR. After Southern blot hybridization 22 of these cases were HPV16-positive, one case was HPV18-positive and 2 cases were not otherwise specified (HPV6, 11 or 33). The dominance of HPV16 in this study is in agreement with previously published results about HPV16 detection in all cervical carcinomas (zur Hausen 1989). In small-cell cervical carcinomas we found only one case with HPV18 infection. This observation is inconsistent with reports that HPV18 is the dominant virus in small-cell cervical cancers (Stoler et al., 1991).

In 8 of the 22 HPV16-positive cases the whole E2 gene, in 9 cases only the DBD or "hinge" region of E2 and in 5 cases neither of them nor TAD could be amplified (table). We suppose that in these 14 cases (64%) with an incomplete or nondetectable E2 gene an integration of the viral DNA into the host cell genome and a deletion or disruption of this gene has occured. The high number of integration events suggests that the integration of HPV DNA into the host genome is an important factor for malignant transformation (Dürst et al 1983; Gissmann et al 1987). The detection of the whole E2 gene in 8 cases could be due either to the episomal form of the virus alone or to the simultaneous prevalence of the integrated and episomal form of HPV.

Additional investigations are necessary for the differentiation between integrated and circular extrachromosomal HPV DNA for example DNA extraction, restriction enzyme digestion andfollowed by 2D gel electrophoresis.

Table 1. Physical state of HPV16 DNA in nonkeratinizing squamous cervical carcinomas. [x]episomal or integrated forms, respectively

No	L1	E2 TAD	E2DBD	E2 hinge	episomal form[x]	integrated form
1	+	+	+	+	+	
2	+	-	+	-		+
3	+	-	-	-		+
4	+	-	-	-		+
5	+	-	-	+		+
6	+	-	+	+		+
7	+	-	+	+		+
8	+	-	-	+		+
9	+	-	-	+		+
10	+	-	-	+		+
11	+	-	-	-		+
12	+	+	+	+	+	
13	+	-	-	+		+
14	+	+	+	+	+	
15	+	-	-	-		+
16	+	+	+	+	+	
17	+	+	+	+	+	
18	+	+	+	+	+	
19	+	+	+	+	+	
20	+	-	-	-		+
21	+	+	+	+	+	
22	+	-	-	+		+

REFERENCES

Dürst M, Kleinheinz A, Hotz M, Gissmann L. The physical state of human papillomavirus type 16 dna in benign and malignant genital tumours. J Gen Virol 1985, 66: 1515-1522

Gissmann L, Dürst M, Oltersdorf T, von Knebel Doeberitz M. Human papillomaviruses and cervical cancer. Cancer Cells 1987, 5: 275-80

Kiene P, Milde-Langosch K, Runkel M, Schulz K, Löning T. A simple and rapid technique to process formalin-fixed, paraffin-embedded tissues for the detection of viruses by the polymerase chain reaction. Virchows Archiv A. Pathological Anatomy and Histopathology 1992, 420: 269-73

Manos MM, Ting Y, Wright DK, Lewis AJ, Broker TR, Wolinsky SM. Use of polymerase chain reaction amplification for the detection of genital human papillomaviruses. Cancer Cells 1989, 7: 209-14

Stoler MH, Mills SE, Gersell DJ, Walker AN. Small cell neuroendocrine carcinoma of the cervix. A human papillomavirus type 18 - associated cancer. Amer J Pathol 1991, 15: 28-32

zur Hausen H. Papillomaviruses as carcinoma viruses. Advanc Viral Oncol 1989, 8: 1-26

PCR in the Field of Bacterial and Fungal Problems

IDENTIFICATION OF TOXOPLASMA-DNA BY POLYMERASE CHAIN REACTION IN PERIPHERAL BLOOD LEUKOCYTES OF PATIENTS WITH SUSPECTED TOXOPLASMOSIS

G. Koch, E.Philipp, A.Handschack, K.Becker and W.A.Müller

Institute for Medical Microbiology, Medical Academy Magdeburg, Germany

INTRODUCTION

PCR has been used to detect Toxoplasma-DNA in peripheral blood leukocytes (PBL) of patients with suspected Toxoplasmosis. Müller et al (1989) could differentiate three groups of patients with circulating Toxoplasma-Antigen (cTgA) in serum samples:

1. Patients with positive detection of cTgA in serum samples over a long period, but without detectable Toxoplasma-specific antibody as tested by Immunfluorescence antibody test (IFAT), Direct agglutination (DA), ISAGA, Toxo-IgM-EIA.

2. Patients with positive detection of cTgA in serum samples for a long period and with low IgG titers (IFAT-IgG 16-256).

3. Patients with positive detection of cTgA in serum samples and with increased antibody titers (IFAT-IgG >1000, Toxo- IgM-EIA or IFAT-IgM positive).

The presence of both cTgA as well as Toxoplasma-DNA could be a sign for parasitemia.

PCR was performed also with samples from patients with acute Toxoplasmosis, where no cTgA was detectable. All samples were obtained from patients with clinical symptoms (lymphadenopathy, encephalitis, chorioretinitis, uveitis, pneumonia) and from pregnant women. In a control group we analysed healthy blood donors with chronical Toxoplasma infection without cTgA.

Methods in DNA Amplification, Edited by
A. Rolfs *et al.*, Plenum Press, New York, 1994

MATERIAL AND METHODS

Isolation of peripheral blood leukocytes from EDTA-blood was performed with PHA-P (SIGMA) according to Ehrlich-Kautzky et al (1991). For cell lysis two methods were tested: A: cell lysis in destilled water at 95°C for 10 min; B: cell lysis in lysis buffer (10 mM TRIS-HCl pH 8,0, 1mM EDTA, 10mM NaCl, 0,5 mg/ml proteinase K, 0,5% SDS).Incubation was at 55°C for 1h and at 37°C for 3h and was followed by DNA extraction with phenol-chloroform-isoamyl-alcohol and ethanol precipitation (Sambrook et al 1989)

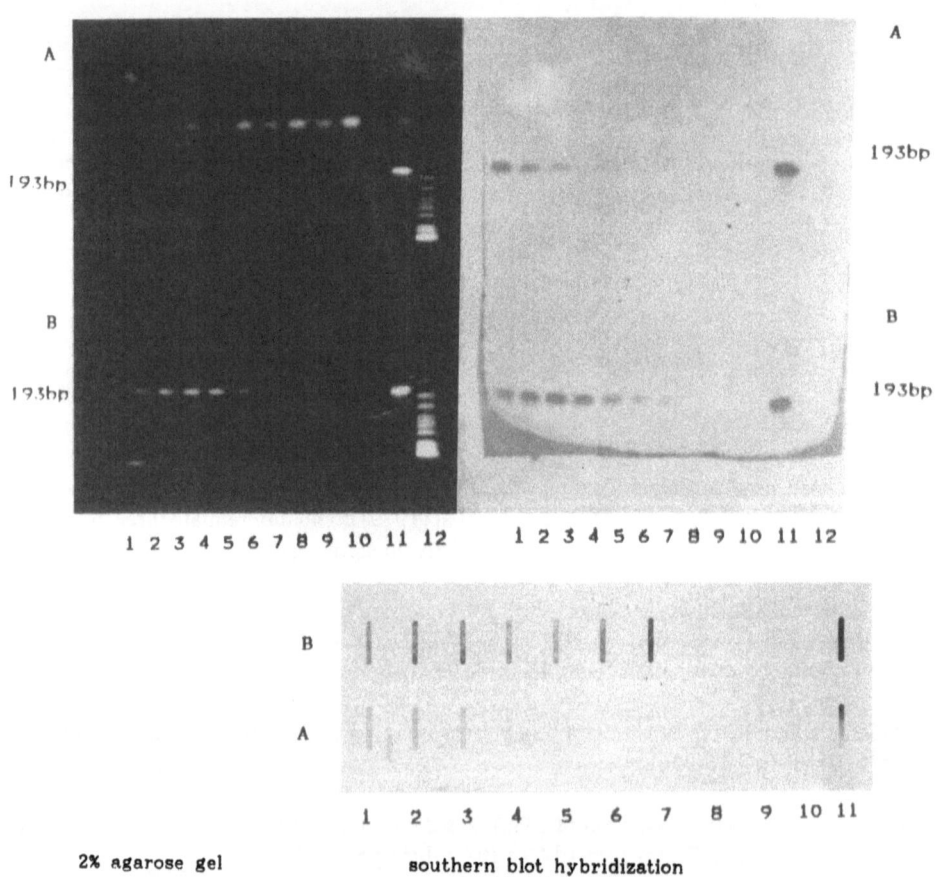

2% agarose gel   southern blot hybridization
                             slot blot hybridization

Figure 1. Amplification of Toxoplasma-DNA (B1-Gene]. Detection sensitivity of tachyzoite dilutions in presence of 1,5 x 10⁶PBL. Demonstration of the results in a 2% agarose gel, southern blot and slot blot procedure. A: cell lysis at 95°C for 10 min; B: DNA extraction phenol-chloroform-isoamylalcohol, ethanol precipitation. lane 1-9: 10^7; 10^6;10^5;10^4;10^3;10^2;10^1;10^0;0 tachyzoites; lane 10: water control; lane 11: positive control; lane 12: DNA molecular weight markers (Boehringer Mannheim DNA weight marker 6)

Toxoplasma-DNA was prepared from Toxoplasma gondii tachyzoites of the RH strain maintained in cell culture P388 (Müller 1985). PCR was performed in a reaction volume of

100µl containing 200 µM dNTP, 1,5 mM MgCl$_2$, 500 mM NaCl, 10 mM TRIS-HCl pH 8,3, 1µM oligonucleotide, 0,001% gelatine, 2,5 U Taq-Polymerase and DNA sample. T. gondii primers described by Burg et al (1989) were used to amplify an 193 bp fragment of the Toxoplasma-B1-Gen.

PCR conditions were denaturation at 94°C for 10 min followed by 40 cycles of denaturation at 94°C for 1 min, annealing at 55°C for 2 min and extension at 72°C for 3 min. Finally the temperature was maintained at 72°C for 10 min to complete the extension.

The following controls were included in every amplification procedure:

1. max. 1mg human DNA of healthy blood donors
2. water control
3. max. 1µg patient DNA + 1-10ng Toxoplasma-DNA
4. 1-10 ng Toxoplasma-DNA

Table 1. Detection of circulating Toxoplasma-Antigen (cTgA) and Toxoplasma-DNA (PCR, B1-Gen) in peripheral blood of 49 patients with suspected Toxoplasmosis and a control group of healthy blood donors with chronic Toxoplasmosis. Low antibody titer means: DA pos., IFAT 16-256, ISAGA nag., IGM-EIA < 0.5; increased antibody titers means: DA pos., IFAT>1000, ISAGA>1600, IgM-EIA>0,9. Clinical symptoms of the patients with suspected Toxoplasmosis: lymphadenopathy, chorioretinitis, encephalitis, uveitis, pneumonia, abortus imminens (gravidity). DNA extraction was done according to method B.

	49 Patients with suspected Toxoplasmosis				control
	38 patients with cTgA			11 patients without cTgA	13 blood donors
negative	antibody titer low	increased		antibody titer increased	antibody titer low
13	23	2		11	13
PCR	PCR	PCR		PCR	PCR
neg pos	neg pos	neg pos		neg pos	neg
9 4	17 6	0 2		7 4	13

PCR products were electrophoresed in 2% agarose gel including DNA molecular weight markers (Boehringer Mannheim 6). The specifity of PCR products was confirmed by slot blot and southern blot hybridization. Hybridization was performed with a digoxigenin labeled internal oligonucleotide (Burg et al 1989) at 55°C for 6h. Hybridization reaction was detected by the anti-digoxigenin- alkaline phosphatase system according to Boehringer Mannheim. Positive results were examined to exclude carry over contaminations by the addition of Uracil DNA Glycosylase (GIBCO BRL). The sensitivity of our detection system was determined by amplifying serial dilutions of Toxoplasma-RH-tachyzoites in the absence and presence of human DNA.

RESULTS AND DISCUSSION

PCR with purified Toxoplasma-DNA results in an 193 bp fragment observed in 2% agarose gel electrophoresis.Ethidium bromide staining allowed the detection of 100-1000 RH tachyzoites. There was up to hundred fold increased sensitivity in slot blot and southern blot hybridization. Figure 1 shows amplification (gel, slot, southern) of several tachyzoite solutions in the presence of $1,5 \times 10^6$ PBL. Preparation method A allowed the detection of 10^5 tachyzoites in slot blot and Southern blot. Extraction of DNA according to method B results in a sensitivity 10-100 tachyzoites in slot blot and southern blot hybridization and should be performed routinely. Table 1 shows the results of the selected patient groups.

For 12 of 38 patients with positive assay of cTgA, Toxoplasma-DNA was detected in PBL; of these patients 4 showed no Toxoplasma specific antibody reaction, 6 had low antibody titers and 2 increased Toxoplasma-antibody titers. In 4 out of 11 patients with acute Toxoplasmosis but without cTgA in serum Toxoplasma-DNA was detectable in PBL. In the control group (13 blood donors) neither cTgA nor Toxoplasma-DNA was observed. Results will be completed by mouse inoculation tests and cell culture. The value of cTgA assay and PCR as a diagnostic tool has to be confirmed in further studies.

REFERENCES

Burg JI, Grover ChM, Pouletty P. Direct and sensitive detetection of pathogenic protozoan Toxoplasma gondii by polymerase chain reaction. J Clin Microbiol 1989, 27: 1787-1792

Ehrlich-Kautzky E, Shinomiya N. Marsh DG. Simplified method for isolation of white cells from whole blood suitable for direct polymerase chain reaction. BioTechniques 1991, 39: 39-40

Müller WA. Stammerhaltung von Toxoplasma gondii in permanenten Mäusemakrophagenkulturen. Angew Parasitol 1985, 26: 221-222

Müller WA, Wohlfahrt A, Koch G. Nachweis von zirkulierendem Toxoplasma-Antigen in Patienten- seren mit monoklonalen Antikörpern in einem Enzymimmuntest. Z Klin Med 1989, 44: 1943- 1946

Sambrook J, Fritsch EF, Maniatis. Molecular cloning- A laboratory manuel, 1989, Cold Spring Harbor Laboratory Press

DETECTION OF *MYCOBACTERIUM TUBERCULOSIS* IN CLINICAL SAMPLES BY POLYMERASE CHAIN REACTION

G.Martinetti Lucchini and M.Altwegg

Institute of Medical Microbiology, University of Zürich, Switzerland

The fact that PCR is not dependent on the growth of organisms represents a considerable advantage regarding sensitivity and rapidity of the detection of slow growing organisms as compared to conventional methods. One of the major problems encountered in the application of the PCR for the detection of mycobacteria in clinical samples is the lysis of bacterial cells. We performed PCR on 43 culture positive clinical samples using primers specific for the *M. tuberculosis* complex developed by Eisenach et al (1990), targeting the multicopy insertion sequence IS6110. DNA suitable for amplification was liberated after a combined treatment which included sonication, incubation in the presence of a detergent and heat. All samples were treated with 0.5 U uracyl-N-glycosylase prior to amplification using dUTP instead of dTTP. The specificity of the PCR was 100%, no amplicons of 123 bp were found when other mycobacteria were detected using conventional methods. In one case we observed a band in a sample from which *M. chelonae* was isolated. However, this band differed in the molecular size as well as in the restriction fragments generated after *Sal*I digestion and it hybridized only very weakly with the labeled probe. The sensitivity of our system was 92% for microscopy positive samples and 60 % for microscopy negative and culture positive samples. We believe that PCR provides reliable, accurate, and rapid results in the diagnosis of *M. tuberculosis* suitable for a routine application in a specialized laboratory. However, it should be combined with cultural procedures.

INTRODUCTION

The laboratory diagnosis of tuberculosis is still based on culture of *Mycobacterium tuberculosis* from secretions and/or tissue samples together with a compatible clinical picture. Because of the low growth rate of mycobacteria, the development of rapid, sensitive and specific tests for the detection of mycobacterial sequences in clinical specimens has been a long-standing goal. The Polymerase chain reaction (PCR) fulfills all these requirements by

Methods in DNA Amplification, Edited by
A. Rolfs *et al.*, Plenum Press, New York, 1994

amplification of a specific DNA fragment located between two short DNA fragments designated as primers.

During the last three years, many systems for the specific amplification of *Mycobacterium tuberculosis* complex with PCR have been described. The primers chosen were based on the nucleotide sequences of genes coding for proteins such as mtp40 (Del Portillo et al., 1991), MPB 64 (Kanedo et al 1990), and MPB 70 (Cousins et al 1992) or on the nucleotide sequences of insertion elements such as IS986 (Kolk et al 1992) or IS6110 (Eisenach et al 1991). We used the system of Eisenach et al. which takes advantage of the presence of 10-15 copies of the repetitive element IS6110 within the chromosome, thus theoretically increasing the sensitivity of the assay.

A problem encountered in the preparation of clinical specimens for PCR analysis is the isolation of mycobacterial DNA. We have developed a rapid and efficient technique which allows routine isolation of mycobacterial DNA suitable for amplification.

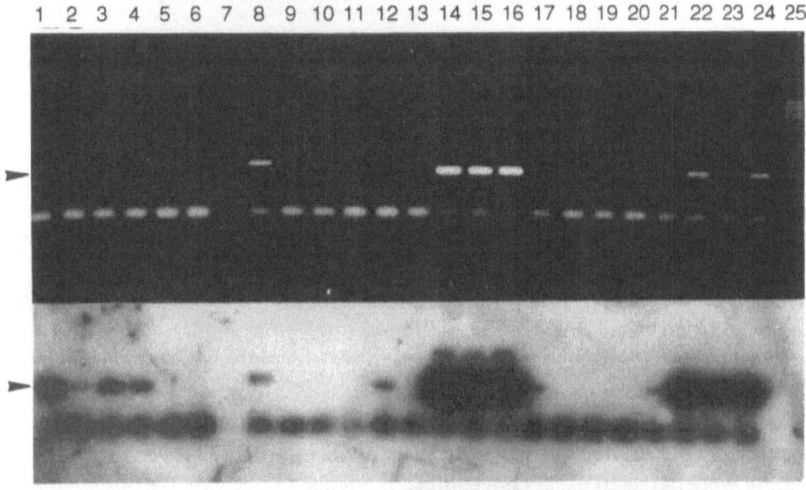

Figure 1. Electrophoretic separation of amplified DNA from clinical specimens by PCR. Ten μl (1/10) of the reaction were electrophoresed on a 3% agarose gel and stained with ethidium bromide (top). The DNA fragments were transferred to a membrane and hybridized with DIG labeled 123 bp fragment (bottom). Lane 1, 2, smear+/culture+ sample, 10^{-1} and 10^{-2} dilutions prior to amplification; lanes 3, 4, smear+/culture- sample, 10^{-1} and 10^{-2} dilutions prior to amplification; lanes 5, 6, smear-/culture+ sample, undiluted and 10^{-1} dilution prior to amplification; lane 7, smear-/culture+ sample, undiluted; lane 8, smear-/culture+ sample containing *M. chelonae* (see Results and Discussion); lanes 9, 10, smear+/culture+ sample, undiluted and 10^{-1} dilution prior to amplification; lanes 11-19, undiluted samples, lane 11, smear-/culture- sample; lane 12, smear-/culture+ sample; lane 13, smear-/culture+ sample containing *M. gordonae*; lane 14, smear+/culture- sample; lanes 15-17, smear+/culture+ samples; lanes 18-19, smear-/culture- samples; lanes 20-21, PCR negative controls; lanes 22-24, PCR positive controls; lane 25, molecular weight marker, pBR322 DNA digested with *Hae*III. The 123 bp target fragment is indicated by the arrow.

MATERIAL AND METHODS

Treatment of clinical samples. Forty three sputum samples were included in this study. Of these, 17 were smear positive and 40 were culture positive for mycobacteria by using conventional solid (1 Loewenstein-Jensen slant, 1 Coletsos slant) as well as liquid media (Bactec). Decontamination was done by incubation with NaOH-sodium citrate-N-acetyl-L-cysteine as described by Kubica et al. (1963).

Two hundred μl of the decontaminated clinical samples were centrifuged for 7 min in a microfuge and pellets resuspended in 500 μl H_2O with 30% (w/v) glass beads (75-150 μm, Sigma St. Louis, MO). Suspensions were sonicated 30 min in a sonication bath (Branson) and incubated briefly at room temperature until the glass beads had settled to the bottom of the tube. 150 μl of the supernatant were transferred to a new tube and mixed with 50 μl lysis buffer to a final concentration of 1% Triton X100, 10 mM Tris-HCl pH 8.0, and 1 mM EDTA and incubated at 95°C for 30 min. Tubes were centrifuged for 7 min and 10 μl of the supernatants were used for amplification.

PCR. The set of primers used were those described by Eisenach et al (1991). Amplification reactions were performed in 100 ml volumes using recombinant Taq DNA Polymerase (Super Taq, ITC Biotechnology GmbH, Heidelberg, Germany) and reagents according to the manufacturer's instructions.

All amplifications were performed using dUTP instead of dTTP and the final composition of the PCR mix was 10 mM Tris-HCl, pH 9.0, 50 mM KCl, 0.01% (w/v) gelatin, 1.5 mM $MgCl_2$, 0.1% Triton X100, 0.2 mM (each) deoxynucleoside triphosphates (dATP, dCTP, dGTP, dUTP), 50 pmol of each primer, 0.25 U of Taq polymerase and 0.5 U of Uracyl N-Glycosylase (UNG) enzyme (Perkin-Elmer Cetus, Norwalk, Conn.), which was added immediately before the template. A short room-temperature incubation step (10 min) was

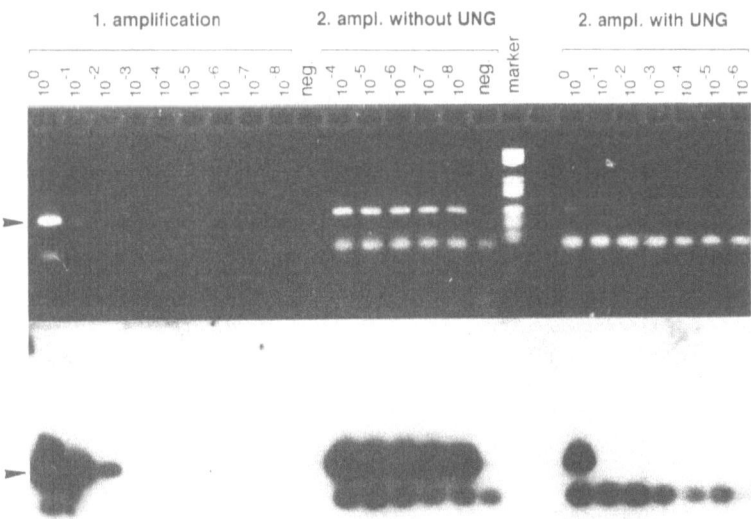

Figure 2. Determination of the effectiveness of the UNG enzyme to prevent carryover of amplicons. Mycobacterial DNA from a smear+/culture+ clinical sample was amplified by a PCR system based on the amplification of a 123 bp fragment of the insertion element IS6110. The resulting product was diluted up to 10^{-8}; 10 μl of each dilution step were loaded on the agarose gel, 10μl of the dilutions 10^{-4}-10^{-8} were reamplified without UNG, and 10 μl of the dilutions 10^0 (undiluted) to 10^{-6} were reamplified with UNG. Top: analysis by agarose gel electrophoresis. Bottom: Southern blot and hybridization with the DIG-11-dUTP- labeled 123 bp probe. The molecular weight marker is the plasmid pBR322 digested with *Hae*III.

performed prior to transfering the tubes to an automated thermal cycler (GeneAmp 9600 Perkin- Elmer Cetus). The samples were denatured at 94°C for 5 min and then 35 cycles (denaturation at 94°C for 1 min, annealing at 68°C for 1 min, extension at 72°C for 1 min) were performed. After a final extension step at 72°C for 10 min, the samples were immediately processed, frozen at -70°C, or kept at 72°C.

Detection of PCR product. PCR products were detected by direct analysis on agarose gels (2.5% NuSieve, FMC, Rockland, ME; 0.5% UltraPure, GIBCO-Bethesda Research Laboratories, Gaithersburg, MD) in TBE buffer (Ausubel et al., 1989). The gels were stained with ethidium bromide and visualized by ultraviolet transillumination. The DNA fragments were then transferred to a nylon membrane (BiodyneA; Pall Biosupport, East Hills, N.Y.) by using the method of Southern (Ausubel et al., 1989). The blots were hybridized with labeled 123-bp fragment produced by PCR. A preparative 3% NuSieve agarose gel (FMC) in TAE buffer (Ausubel et al., 1989) was performed, the 123 bp fragment was extracted and purified using a SpinBind DNA extraction unit (FMC, Rockland, ME) following the instructions of the manufacturer. Thereafter the DNA was reamplified using the same protocol except that the amplification mixture contained 40 nM each of dATP, dCTP and dGTP, 14 nM digoxigenin-11-dUTP (DIG-11-dUTP) and 26 nM dUTP and 40 cycles were performed. The hybridization was done at 63°C, in 5X SSC (1X SSC is 0.15 M NaCl, 0.015 M Na_3 citrate, pH 7.0), 0.1 % Lauroylsarcosine, 0.02% SDS, and 2% Blocking reagent (Boehringer-Mannheim, Mannhein, Germany). Hybrids were detected with the DIG Luminescent Detection Kit following the instructions provided by the manufacturer (Boehringer Mannheim).

Table 1. Sensitivity and specificity of the detection of *Mycobacterium tuberculosis* in clinical samples. EtBr, after ethidium bromide staining; Hyb, after hybridization with DIG labeled fragment.

Mycobacterium tuberculosis		EtBr+	Hyb+	
smear+/cultivation+	n = 10	7	9	sensitivity 90 %
smear+/cultivation-	n = 3	3	3	sensitivity 100 %
smear-/cultivation+	n = 5	0	3	sensitivity 60 %
Nontuberculosis mycobacteria				
smear+/cultivation+	n = 4	0	0	specificity 100 %
smear-/cultivation+	n = 21	0	0	specificity 100 %

RESULTS AND DISCUSSION

Preparation of clinical samples. Isolation of mycobacterial DNA suitable for amplification represents one of the major problems for PCR diagnostics. Lysis of bacterial cell with proteolytic enzymes is often inefficient because of their cell wall composition. Our major goal was to isolate DNA which permits direct PCR amplification, thereby omitting organic extraction and other tedious procedures which involve numerous steps with possible loss of material. These factors are of particular importance in a routine laboratory, where many samples will be processed simultaneously. In the first step of our procedure, cells were damaged by sonication in the presence of glass beads. Cells were then lysed with TE-Triton X100 at high temperature. Although not systematically optimized, this procedure resulted in a good performance of the test (see below). DNA isolation procedures reported in the literature are controversial. Sritharan and Barker Jr (1991) compared different lysis methods and demonstrated that using the TE-Triton X100 boiling procedure alone, they were able to detect one single cell of cultured *M. tuberculosis* after hybridization. However Buck et al (1992) reported that sonication of clinical samples produced the most promising results as compared to other methods including boiling in non ionic detergents as Tween 20 and Triton X100. In previous experiments (data not shown) we combined sonication and heat and obtained a higher sensitivity when clinical samples were sonicated in the presence of glass beads.

Amplification and detection. DNA isolated from respiratory samples was amplified with primers located within the insertion sequence IS6110 resulting in an amplified fragment of 123 bp (Figure 1). Results of detection using ethidium bromide staining and hybridization are summarized in Table 1. Hybridization with the DIG-labeled fragment increases the sensitivity of the detection by a factor 100 as compared to ethidium bromide staining (data not shown). We analyzed 13 smear-positive samples, 10 of which were also culture-positive for M. tuberculosis. Twelve were positive after hybridization resulting in a sensitivity of 92%. For smear-negative, culture-positive samples the sensitivity decreased to 60%. The sensitivity values that we obtained with our method can be compared with those published using the same set of primers (Eisenach et al 1991) or other targets such as IS986 (Kolk et al 1992) or the gene coding for the 32-KD Protein (Soini et al 1992). However, it is difficult to compare results obtained by different groups, since not only extraction and amplification procedures were different but also the methodology and media used for culture. It should be pointed out that most of published techniques were evaluated with smear-positive clinical samples and compared to not always satisfactory cultural procedures. The reported sensitivities were similar irrespective of whether the target genes were present as single copy or as multiple copies within the genome.

The specificity of our assay was tested by amplification of respiratory samples containing nontuberculous mycobacteria which have been identified at the species level using conventional methods. The 123 bp fragment was found in none of the 25 clinical specimens analyzed and, therefore, the specificity was 100%. Nevertheless a fragment of approximately 140 bp was amplified in a clinical sample containing *M. chelonae* (Figure 1). Restriction enzyme analysis showed that this fragment was not digested with SalI, whereas digestion of *M. tuberculosis* yielded two fragments of 55 and 68 bp, respectively (data not shown). The hybridization of the 140 bp fragment with the DIG labeled 123 bp fragment of *M. tuberculosis* was very weak, indicating the presence of nonhomologous sequences. Since we used the complete fragment as probe, the low signal could be derived from the hybridization of primer sequences.

UNG-treatment. We assessed the effectiveness of the UNG enzyme to prevent carry-over of amplicons. One amplified fragment was diluted up to 10^{-8} and reamplified using the same protocol as for clinical specimens. Dilutions 10^{-5} to 10^{-8} were also reamplified without addition of UNG. Results obtained after staining in ethidium bromide and hybridization are shown in figure 2. A positive amplification of the UNG treated samples was obtained only with the highest amplicon concentration (undiluted), whereas reamplification of the diluted fragments without UNG yielded a visible fragment until dilution 10^{-8}. Recently Epsy et al. (1993) reported that small amplicons (á 98 bp) are not affected by this enzymatic treatment and they suggested that for optimal performance of the UNG-carryover prevention system an amplicon size of at least 150 bp should be chosen. Although the size of our amplicon was only 123 bp, 0.5 U UNG were able to efficiently destroy higher amounts of DNA than those that are likely to be carried over by good laboratory practice.

In summary, the sonication and heat procedure described here appears to be a simple, rapid method for treating clinical specimens containing *M. tuberculosis* and to release DNA for amplification. Despite the relatively low number of specimens analyzed we believe that this system offers a reliable method for the detection of *M. tuberculosis* in clinical samples. Further studies especially involving more smear-negative, culture-positive specimens are necessary to fully evaluate this procedure.

ACKNOWLEDGEMENTS

We would like to express our gratitude to J. Mosimann and A. Descombes (Institut Microbion, Lausanne, Switzerland) as well as M. Grubenmann (Laborgemeinschaft 1, Zürich) for providing clinical specimens.

REFERENCES

Ausubel FM, Brent R, Kingston RE, Moore DD, Seidman JG, Smith JA and Struhl K. Current protocols in molecular biology. John Wiley and Sons, Chichester, United Kindom 1989

Buck GE, O'Hara LC and Summersgill JT. Rapid, simple method for treating clinical specimens containing *Mycobacterium tuberculosis* to remove DNA for polymerase chain reaction. J Clin Microbiol 1992, 30:1331-1334

Cousins DV, Wilton SD, Francis BR and Glow BL. Use of polymerase chain reaction for rapid diagnosis of tuberculosis. J Clin Microbiol 1992, 30:255-258

Del Portillo P, Murillo LA and Patarroyo ME. Amplification of a species-specific DNA fragment of *Mycobacterium tuberculosis* and its possible use in diagnosis. J Clin Microbiol 1991, 29:2163-2168

Eisenach KD, Cave MD, Bates JH and Crawford JT. Polymerase chain reaction amplification for a repetitive DNA sequence specific for *Mycobacterium tuberculosis*. J Infect Dis 1990, 161:977-981

Eisenach KD, Sifford MD, Cave DM, Bates JH and Crawford JT. Detection of *Mycobacterium tuberculosis* in sputum samples using a polymerase chain reaction. Am Rev Respir Dis 1991, 144: 1160-1163

Espy MJ, Smith TF and Persing DH. Effect of amplicon size and nucleotide composition on inactivation by isopsoralen (IP) and uracyl-N-glycosylase (UNG) after amplification by polymerase chain reaction (PCR). Abstr. 93rd General Meeting of the American Society for Microbiology, 1993, 485

Kanedo K, Onodera O, Miyatake T and Tsuji S. Rapid diagnosis of tuberculous meningitis by polymerase chain reaction (PCR). Neurology, 1990, 40, 1617-1618

Kolk AHJ, Schuitema ARJ, Kuijper S, van Leeuwen J, Hermans PWM, van Embden JDA and Hartskeerl RA. Detection of *Mycobacterium tuberculosis* in clinical samples by using Polymerase chain reaction and a nonradioactive detection system. J Clin Microbiol 1992, 30:2567-2575

Kubica GPW, Dye E, Cohn ML and Middlebrook G. Sputum digestion and decontamination with N-acetyl-L-cysteine-sodium hydroxide for culture of mycobacteria. Am Rev Respir Dis 1963, 87:775-779

Soini H, Skurnik M, Liippo K, Tala E and Viljanen MK. Detection and identification of mycobacteria by amplification of a segment of the gene coding for the 32-kilodalton protein. J Clin Microbiol 1992, 30:2025-2028

Sritharan V, and Barker Jr RH. A simple method for diagnosing *M. tuberculosis* infection in clinical samples using PCR. Mol Cell Probes 1991, 5:385-395

TUBERCULOSIS IN HIV INFECTED PATIENTS DETECTED BY PCR: A COMPARISON WITH CLINICAL DATA

U.Thums[1], H.D.Brede[1], H.Rübsamen-Waigmann[1], R.Brodt[2], E.B.Helm[2], C.Schneider[3], V.Brade[3]

[1]Chemotherapeutisches Forschungsinstitut Georg-Speyer-Haus, Frankfurt/Main; [2]Zentrum Innere Medizin Universitätskliniken Frankfurt; [3]Abt. Mikrobiologie im Zentrum für Hygiene der Universitätskliniken Frankfurt/Main, Germany

INTRODUCTION

Various polymerase chain reaction (PCR) assays have been described for the rapid identification of mycobacteria in clinical specimens. To assess the value of the assays in routine laboratory work the results obtained by PCR were compared with those obtained by standard microbiological methods and clinical data. 56 sputum specimens were collected for investigation of mycobacterial infection in HIV-infected patients, 73 from patients without HIV-infection. In a substantial number of HIV patients with tuberculosis both chest x-ray and the clinical course were found to be unusual, in concordance with published data. Rapid detection of mycobacteria by laboratory methods is therefore of great importance for the care of HIV-infected patients. Isolation of mycobacteria from clinical samples takes up to 8 weeks for culture on solid medium and up to 20 days by the BACTEC system, which is based on the measurement of carbon dioxide released by bacteria during growth in liquid medium. It has often been reported that amplification of mycobacterial DNA by use of the polymerase chain reaction (PCR) technique permits rapid detection of mycobacteria directly in clinical samples.

Specimens were tested for the presence of *Mycobacterium tuberculosis (M.tb)* complex and atypical mycobacteria in two assays, one based on amplification of the 65 kDa gene and the other on the 16S RNA-gene. For the 129 samples that did not contain inhibitors of the amplification reaction PCR findings correlated well with bacteriological and clinical data in 106 (82,2%). One PCR result turned out to be false negative, 32 to be false positives. Evaluation of clinical data of these false positives demonstrated that 12 of 30 patients had received tuberculostatic treatment, and in several cases a later culture turned positive for *M. tb* or tuberculosis was confirmed by authopsy. This study confirms the potential of DNA amplification for early diagnosis of mycobacterial infections. The results demonstrate that a positive PCR may indicate true infection and that comparison with culture results is not

Methods in DNA Amplification, Edited by
A. Rolfs *et al.*, Plenum Press, New York, 1994

sufficient in evaluation of PCR for tuberculosis diagnosis - clinical and pathological data must be included.

Clinical specimens

All specimens were sputum specimens, cleaned with the N-Acetyl-L-Cystein-NaOH-method as described in DIN 58 943. 56 specimens were obtained from 21 HIV patients (19 CDC IV b-e, 3 CDC IV a) with a high index of suspicion for tuberculosis who were inpatients at the University Hospital at Frankfurt/Main, Germany. Each of these patients was positive by PCR in at least one specimen. 73 specimens were obtained from 67 non- HIV-infected patients who were suspected to have tuberculosis.

Sample treatment

Specimens were aliquoted in 5 parts, 3 for cultivation, one on reserve and one for treatment with sodium hydroxide lysis and phenol extraction as previously described (Brisson-Noel et al, 1989; Maniatis et al, 1984). Samples containing PCR inhibitors or giving false-negative results were further treated as follows before being retested with the 65 kD and 16S RNA-assay: 50 to 100µl of sample were lysed by guanidium thiocyanate and DNA was purified on silica particles (Boom et al, 1990). Controls with a Mc.Farland-standard containing 10-100 mycobacteria and with only aqua dest. were included in each experiment.

DNA amplification

Amplification by polymerase chain reaction was done as discribed (Hance et al, 1989; Wilton et al, 1992; Böddinghaus et al, 1990) with 10µl of DNA extracted from the clinical sample. For thermal cycles and primers for both nested-PCR-assays see table 1. Controls with purified M. tuberculosis DNA (300 fg = ca. 30 bacteria) and without DNA were included in each experiment.

Samples were considered positive for tuberculosis when repeatedly positive in the two independent experiments with either the 65 kD antigen-gene or the 16S RNA-gene assay.

Labelling of probes and hybridisation experiments

Oligonucleotids (65 kD, see table 1) were radiolabelled at the 5' by use of polynucleotide kinase (Boehringer Mannheim,Germany) and (γ^{32}P)-ATP (specific activity 3000 Ci/mmol, Amersham).

The TB1-TB2 assay amplification products were detected with an oligonucleotide probe specific for *M.tuberculosis* complex (TB3) (or with a probe specific for *M.avium-intracellulare*-complex, an other MOTT, data not shown) by liquid hybridisation, separation by gel-electrophresis and documentation on x-ray. The use of ^{32}P-labelled probes and nested-PCR-assay led to identical results.

Table 1. Primers and amplification protocol

---

65 kD AG-gene	16S rRNA-gene

specific for genus
TB1: 5′ GAGATCGAGCTGGAGCC 3′
TB2: 5′ AGCTGCCCAAAGGTGTT 3′

specific for genus:
Mycgen-F: 5′ AGAGTTTGATCCTGGCTGAG 3′
Mycgen-R: 5′ TGCACACAGGCCACAAGGGA 3′

35 cycles: 95°C, 60 sec
            55°C, 60 sec
            72°C, 60 sec

10µl of genus-specific product

specific for *M.tb*-species
TB 3: 5′ GCGGCATCGAAACCGTG 3′ (*)
TB 4: 5′ CGAAATCGCTGCGGCCG 3′

specific for *M.tb*-species:
TB-1F: 5′ GAACAATCCGGAGTTGACAA 3′
TB-1R: 5′ AGCACGCTGTCAATCATGTA 3′

35 cycles: 95°C, 50 sec
            74°C, 120 sec

Single-tube-amplification:
45 cycles: 94°C, 30 sec
            60°C, 180 sec
            75°C, 180 sec

(*) also used as hybridization-oligonucleotide

RESULTS

Comparison of amplification results with bacteriological data

129 specimens were tested by PCR. All of these were also sent for culture on Löwenstein-Jenson solid medium and/or MB-Check-System (Roche). 106 specimens gave the same result wether tested by DNA amplification or culture. These specimens were either positive for *M.tb*-complex (26), positive for atypical mycobacteria (2 *M.avium*), or negative samples (79), including 1 HIV-infected patient without any indication of tuberculous infection. Discrepancies were found in 33 specimens - 32 were culture negative and PCR-positive, 1 was culture positive and PCR negative.

When samples containing inhibitors or otherwise giving false-negative results were treated with guanidium thiocyanate silica, positive results were obtained for two 2 of 3 samples that were culture-positive and PCR-negative with the sodium hydroxide phenol protocol.

Table 2. Comparison of amplification results and bacteriological data. (*) 6/18 received tuberculostatic treatment (1 no data available); (**) 6/12 received tuberculostatic treatment (1 no data available)

| | Nested-PCR-Amplification | | | | | | |
| | positive | | | negative | | |
Culture	HIV	non- HIV	total	HIV	non- HIV	total
positive	14	4	18	1	0	1
negative	19(*)	13(**)	32	22	56	78
total	33	17	50	23	56	79

Comparison of DNA amplification results with clinical criteria

32 specimens were positive by PCR analysis but culture negative. PCR-positive findings were examined in relation to clinical criteria for mycobacterial infetions. 12 of these samples came from 2 patients with bacteriological confirmation of mycobacterial infection in other specimens

5 patients had bacteriological confirmation of the PCR findings in a later or an earlier sputum specimen, one patient had confirmation by biopsy and in 6 cases there was clinical improvement with antituberculous treatment (Table 3). In some patients several different samples were PCR positive, which strengthened the evidence for tuberculous infection. In other patients various symptoms of mycobacterial infection were seen.

In concordance with published data in a substantial number of HIV-patients with tuberculosis both chest x-ray and clinical course were found to be atypical:

Nonspecific constitutional symptoms were most common, including lost of weight (n = 18, 86%), fever (n = 20, 95%), cough (n = 20, 95%), nonspecific chest x-ray with diffuse infiltrate (n = 10, 48%), pulmonary lymphadenopathy (n = 7, 33%) and even normal-appearing chest x-ray (n = 7, 33%). 10 patiens also had extrapulmonary manifestations.

All the patients with atypical mycobacteria had clinical signs suggestive of a mycobacterial infection.

DISCUSSION

Our investigation of how well PCR works in the laboraory diagnosis of tuberculosis shows that the DNA amplification technique is more sensitive than bacteriological culture. Of the 129 specimens that were free of amplification-inhibitors 106 (82,2%) gave results that accorded well with bacteriological samples, 118 (91,5%) with bacteriological samples and clinical data (12 positive samples from patients with later positive culture, tuberculostatic treatment or dianosis by authopsy but negative culture). Of 41 specimens from patients with a definite diagnosis of tuberulosis, 40 were positive by DNA amplification and only 28 by culture. 12 of 41 mycobacteria-positive patients were identified by DNA amplification but not by microbiological methods (table 2). The false-negative results could be due to: first, the presence of inhibitors not eliminated by guanidinium-thiocyanate-teatment and not detected by the control amplification; second, non-homogeneous distribution of bacteria in the specimen so that the fraction tested does not contain mycobacteria; or third, low numbers of bacilli in the specimen (sensitivity tested by dilution of Mc.-Farland standards down to 1-10 mycobacteria per ml), which decreases the probability of presence of mycobacteria after aliquotation in the fraction analysed by PCR.

Thus to minimise the possibility of obtaining false-negative results DNA amplification has to be tested in several specimens for each patient which is also true of culture methods. Inhibitors of the amplification reaction were detected in 3 PCR-positive samples. They were not identified, but could result from residual traces of sodium dodecyl sulphate or phenol, which are known to be potent inhibitors of Taq polymerase. They might also be patient-associated.

Table 3: Comparison of culture and PCR results in 21 HIV-infected patients

pat.	positive culture	positive PCR	clinics	tuberculostatic therapy
1	0/1	1/1 M.tb	post mortem M.tbc in liver/lung	no
2	0/1	1/1 M.tb	culture pos. in a subsequent sputum-sample	yes
3	0/1	1/1 M.tb	no indication of tuberculosis	no
4	0/1	1/1 M.tb	a subsequent blood-culture positive	no
5	0/1	1/1 M.tb	no indication of tuberculosis	no
6	0/1	1/1 M.tb	culture pos. in a prior specimen	yes
7	0/1	1/1 MOTT	a subseqeunt blood-culture MAI-positive	no
8	0/1	1/1 M.tb	Fever, lost of weight,inconspicuous x-ray	no
9	5/7 M.tb	7/7 M.tb	disseminated tuberculosis	yes
10	0/3	2/3 M.tb	culture pos. in a subsequent sputum-sample	yes
11	0/5	1/5 M.tb	recurrent bacterial pneumonia	no
12	1/2 M.tb	1/2 M.tb	disseminated tuberculosis	yes
13	0/3	1/3 M.tb	positive M.tb culture of lymphe-node	yes
14	1/2 MAI	1/2 M.avium	disseminated MAI-infection	?
15	0/5	1/5 M.avium	a subseqeunt culture MAI-positive	no
16	4/5 M.tb	5/5 M.tb	tuberculosis	yes
17	1/2 M.tb	2/2 M.tb	disseminated tuberculosis	yes
18	1/3	2/3 M.tb	tuberculosis	yes
19	1/3 MAI	0/3	MAI in blood-culture	?
20	1/3 M.tb	1/3 M.tb+MAI	MAI demonstrated in liver	yes
21	0/5	1/5 M.tb	no indication of tuberculosis	no

For some samples results with one amplification assay differed from that obtained with the other. The lesson is that two amplification assays should be done in parallel. In conclusion, nucleic acid amplification (and specific probe-hybridisation) is a reliable method for the early diagnosis of mycobacterial infections. This method would be especially useful in the diagnosis of tuberculosis in HIV-infected patients since infection is usually disseminated in these cases and can follow an unusual clinical course and have an atypical chest x-ray. Early detection of mycobacteria by laboratory methods is therefore of great importance for the care of HIV-infected patients. In view of possible contaminations, inhibitors and other problems of PCR in direct demonstration of microorganisms proof of germs in clinical specimens this method should be reserved to experienced laboratories. PCR has yet to be established as a routine laboratory method, and more prospective studies on the correlations between PCR, bacteriological methods and clinical infections need to be performed.

ACKNOWLEDGMENT

This study was supported by Bundesministerium für Gesundheit, FRG. We thank Mrs. Landersz for her help in laboratory.

REFERENCES

Böddinghaus B, Rogall T, Flohr T, Blocker H, Bottger EC. Detection and identification of mycobacteria by amplificationof rRNA. J Clin Microbiol 1990, 28: 1751-1759

Boom R, Sol CJ, Salimans MM, Jansen CL, Wertheim-van Dillen PN, van der Noordaa J. Rapid and simple method for purification of nucleic acids. J Clin Microbiol 1990, 28: 495-503

Brisson-Noel A, Lecossier D, Nassif X, Gicquel B, Levy-Frebault V, Hance AJ. Rapid diagnosis of tuberculosis by amplification of mycobycterial DNA in clinical samples. Lancet 1989; II: 1069-1071

Hance AJ, Grandchamp B, Levy-Frebault V, Lecossier D, Rauzier J, Bocart D, Gicquel B. Detection and identification of mycobacteria by amplification of mycobacterial DNA. Molec Microb 1989, 3: 843-849

Maniatis T., Fritsch EF., Sambrook J.: Molecular cloninig, a laboratory manual. New York: Cold Spring Harbor Press, 1984

Wilton S, Cousins D. Detection and identification of multiple mycobacterial pathogens by DNA amplification in a single tube. PCR Methods and Applications 1992, 1: 269-273

DETECTION OF *BORRELIA BURGDORFERI* IN HUMAN SKIN BIOPSIES BY A NESTED POLYMERASE CHAIN REACTION

S. E. Moter[1], R. Wallich[2], M. M. Simon[3] and M. D. Kramer[1]

[1]Institute for Immunology, [2]German Cancer Research Center, Heidelberg, [3]Max-Planck-Institute for Immunobiology, Freiburg, Germany

INTRODUCTION

Borrelia burgdorferi (*B. burgdorferi*) is the etiologic agent of Lyme disease (also termed: Lyme borreliosis), a worldwide endemic infectious disease. The bacterium *B. burgdorferi* belongs to the order Spirochaetales. It is transmitted to animals and humans by ticks of the genus *Ixodes*. In Europe *Ixodes ricinus*, the sheep tick, is the main vector for transmission of *B. burgdorferi*. Infection of humans with *B. burgdorferi* may affect different organ systems causing a variety of clinical symptoms (Table 1). Following the tick-bite the disease can progress from an early stage to a stage of generalized infection and after months to years of duration to a late chronic stage (Steere 1989).

In the early stage of the disease, i.e. at the stage of the local infection of the skin at the site of the tick bite, the diagnosis is based on clinical findings. The typical skin manifestation of the early stage is erythema migrans, which however is seen in only about 50% of the patients a few days after the tick bite. Skin manifestations in Lyme disease occur also during late stages of the infection. The skin manifestation of the late stage is acrodermatitis chronica atrophicans, which may develop years after the onset of the disease and presents as an atrophy of the skin.

After hematogenous spread of B. *burgdorferi* and involvement of internal organs the clinical symptoms are often non-specific or may even mimic other diseases. In these cases the diagnosis has to be supported by laboratory methods. Current laboratory diagnosis, however, is hampered by a number of limitations (summarized by: Schwan et al 1991). *In vitro* cultivation of *Borrelia* is difficult because of the paucity of organisms in infected tissues and is a time-consuming procedure - due to its low doubling-time. Staining with specific antibodies or unspecific silver staining is difficult to evaluate and is not a routine method. Serologic tests exhibit a limited sensitivity and specificity. An alternative method for pathogen detection is the amplification of *B. burgdorferi* gene sequences by the polymerase chain reaction. Therefore we have developed a PCR for the detection of *B. burgdorferi* in human skin.

Methods in DNA Amplification, Edited by
A. Rolfs *et al.*, Plenum Press, New York, 1994

Table 1. Manifestations of human Lyme disease

time course	organ system	symptoms
early stage (1 - 8 weeks)	local infection of the skin at the site of the tick bite	erythema (chronicum) migrans lymphadenosis benigna cutis
end of early stage	hematogenous generalization	general symptoms (headache, fever, arthralgia) lymphadenopathy
late stage (2 - 12 months)	organ manifestations: joints nervous system heart liver	mono-or oligoarthritis meningopolyneuritis, meningoradiculitis carditis acute hepatitis
chronic stage (1 - 10 years)	persistent organ manifestations joints nervous system heart skin	polyarthritis polyneuritis, progressive encephalomyelitis pancarditis acrodermatitis chronica atrophicans

MATERIALS AND METHODS

Patients

Punch biopsies were obtained from patients with skin manifestations (erythema migrans, acrodermatitis chronica atrophicans, scleroderma-like manifestations) suggestive of Lyme disease. Specimens were sent to the Immunopathology Laboratory (Institute for Immunology, University of Heidelberg) by dermatologists of different hospitals in Germany. As negative controls, biopsies from a person with no signs of localized or generalized skin manifestations were included. Biopsies were frozen in liquid nitrogen and stored at -80°C until they were tested by PCR.

Sample preparation

Ten cryostat sections (14 µm each) of each biopsy were placed in a reaction tube containing 100µl of PCR reaction buffer, 0.5% (v/v) NP 40, 0.5% (v/v) Tween-20 and 400µg/ml proteinase K. The suspension was incubated overnight at 55°C followed by 10 min at 95°C and placed on ice. Samples were briefly centrifuged (10,000 x g) and 25 µl of the preparation were added to a PCR reaction. For each biopsy the sample preparation and the PCR were performed in triplicate.

PCR

As a target for PCR amplification the gene coding for the outer surface protein A (ospA) of *B. burgdorferi* was selected (Figure 1). The ospA gene is located on a 49 kb linear plasmid and is present in a single or low copy number (Howe et al 1985, Hinnebusch, Barbour 1992).
In order to obtain a high sensitivity, which is necessary for the detection of *B. burgdorferi* gene sequences in naturally infected tissues, a nested PCR protocol was used. In the first round of PCR cycling, the outer primers prZS7/31-1 (5'-GGGAATAGGTCTAATATTAGCC-3') and OspA-5 (5'-CACTAATTGTTAAAGTGGAAGT-3') were used to amplify a 662 bp fragment of the ospA gene specific for *B. burgdorferi*. In the second round of PCR cycling the inner primers OspA-6 (5'-GCAAAATGTTAGCAGCCTTGACG-3') and OspA-8 (5'-CTGTGTATTCAAGTCTGGTTCC-3') were used to amplify a 393 bp fragment.

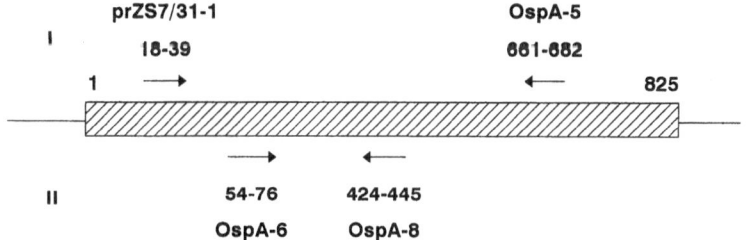

Figure 1. Map of the *B. burgdorferi* ospA gene. The location (in nucleotides) and orientation (indicated by arrows) of primers in the outer PCR (I) and inner PCR (II) are shown.

In the same reaction tube primers for the human β_2-microglobulin gene (Güssow et al. 1987) were used as an internal control. In the outer PCR using the primers β_2-m-1 (5'-ACCCCCACTGAAAAAGATGAGTAT-3') and β_2-m-2 (5'-ATGATGCTGCTTACATGT CTCGAT-3') a 815 bp fragment was amplified. In the inner PCR using the primers β_2-m-2 (see above) and β_2-m-3 (5'-CATGTGACTTTGTCACAGCCCAAG-3') a 677 bp segment was amplified. Amplification of *B. burgdorferi*-specific target sequences was carried out in a total volume of 50 µl. The reaction mixture contained 25 µl of the proteinase K digest of the biopsies, PCR reaction buffer (50 mM KCl, 10 mM Tris-HCl pH 8.3, 1.5 mM $MgCl_2$, 0.01 % (v/v) gelatine), 100 µM dNTP, 0.25 µM of each outer primer (prZS7/31-1, OspA-5, β_2-m-1 and ß₂-m-2) and 1.5 U of *Taq*-DNA polymerase (Boehringer Mannheim). The final concentration of magnesium chloride was 2.6 mM. The reaction was overlaid with 50 µl mineral oil to prevent evaporation. The amplification reaction was carried out for 30 cycles in a DNA thermal cycler (Biomed) with the following cycling profile: denaturation at 94°C for 90 s, primer annealing at 45°C for 60 s, and extension at 74°C for 120 s. Low stringency conditions were used for optimal amplification. After the last cycle, 5 µl of the reaction mixture were added to a new PCR mixture containing 0.125 µM of each inner primer (OspA-6, OspA-8, ß₂-m-2 and ß₂-m-3). The final concentration of magnesium chloride was 1.5 mM. The target sequence was subjected to 25 additional cycles. The annealing temperature was raised to 55°C, in order to achieve high stringency conditions and to exclude amplification of non-target sequences.

PCR products were analysed by gel electrophoresis using a 1.5% agarose gel stained with ethidium bromide and photodocumented.

A panel of positive and negative controls was included in each experiment to control for contaminations, which would result in false positive results, and for false-negative amplification results. Positive controls included *B. burgdorferi* and human target DNA in concentrations just beyond the detection limit of the assay. Furthermore, tissues from experimentally infected mice were processed in parallel. Negative controls included preparations which were processed like sample specimens during sample preparation and PCR but lacked both target DNA or tissue-derived samples. In each experiment a reaction mixture containing TE-buffer served as a negative control. A sample of an uninfected human skin biopsy was processed in parallel.

To avoid contaminations, the set-up of the PCR mixtures, the sample preparation and the analysis of amplification products took place in separate working areas. Pipettes with disposable plungers were used for sample handling and amplification products of the outer PCR. During the transfer of the outer PCR product into the reaction setup for the inner PCR, fresh gloves were used for the handling of each biopsy.

RESULTS

The detection limit of the PCR was 1 fg, when genomic *B.burgdorferi* (strain Z57) DNA was used. When a dilution series of cultured spirochetes ranging from 10^8 to 1 organism was amplified, the detection limit corresponded to 1 spirochete. No amplification products were observed, when 1 pg of DNA of several other species of spirochetes, *i.e. Borrelia hermsii, B. parkeri, B. turicatae, Treponema pallidum* and *T. denticola*, was tested. The findings indicated the absence of cross-reactions of *B. burgdorferi*-specific primers with these species. No amplification products were observed when a proteinase K digest of uninfected human skin or human DNA was tested.

In our hands a simple proteinase K digestion of samples has turned out to be sufficient for the preparation of human skin biopsy specimens as compared to sample preparation by boiling; alkaline lysis or isolation of DNA using phenol/chloroform extraction.

12 human skin biopsies from patients with erythema migrans, acrodermatitis chronica atrophicans or scleroderma-like skin manifestations were tested by PCR. The results are listed in table 2. Samples from 3 out of 5 patients with erythema migrans resulted in positive PCR signals. 2 out of 6 acrodermatitis patients were positive in the PCR. Positive PCR results were obtained from patients who showed an elevated IgG serum antibody titer as well as from patients showing no such elevated antibody titers. In figure 2, examples of a PCR amplification

using samples from two patients with erythema migrans are shown.

Interestingly, one skin biopsy of a patient suffering from scleroderma-like skin manifestations, which could not be diagnosed as a typical skin manifestation caused by *B. burgdorferi*, also showed positive PCR results. Together with the finding of an elevated anti-*B. burgdorferi* IgG titer in the patient's serum, it was decided to commence antibiotic therapy. The assumption that the observed scleroderma-like manifestations were due to a *B. burgdorferi* infection was then corroborated by the fact that the patient's skin symptoms as well as his accompanying neurological symptoms (polyneuropathy) markedly improved after he had received antibiotic therapy (2g/d ceftriaxone for 2 weeks).

Table 2. Results of PCR for the detection of B. burgdorferi DNA sequences in human skin biopsies. ACA: acrodermatitis chronica atrophicans, EM: erythema migrans, EM?: suspected EM, f: female, IgG: results of serologic testing in a *B. burgdorferi*-specific enzyme-linked immunosorbent assay for the detection of IgG antibodies, m: male, nd: not determined, nk: not known, Sclerod.-like: scleroderma-like skin manifestation.

biopsy	patient			PCR	anti -*B. burgdorferi*
no	sex	age	diagnosis	result	IgG in serum
1	m	66	EM	-	nd
2	f	53	EM	+	-
3	m	53	EM?	-	nd
4	f	51	EM?	+	-
5	m	23	EM?	+	-
6	f	48	ACA	-	nd
7	f	46	ACA	-	+/-
8	f	89	ACA	-	++
9	f	64	ACA	-	nd
10	nk	48	ACA	+	+/-
11	m	73	ACA	+	++
12	m	62	Sclerod.-like	+	++

CONCLUSION

Our protocol for nested PCR provides a sensitive and specific method for the detection of *B. burgdorferi* DNA in naturally infected skin. We have designed primers which are capable of detecting all known ospA subtypes of different *B. burgdorferi* isolates. This is of great inportance for PCR amplification in unknown samples, since a considerable genetic heterogeneity among ospA genes of of *B. burgdorferi* isolates is known (Wallich et al 1992). In our hands

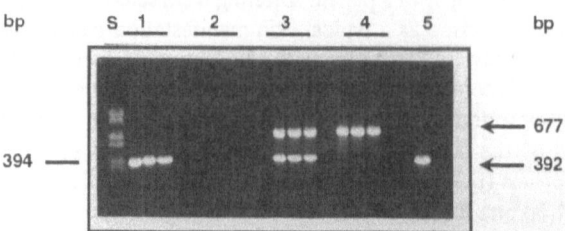

Figure 2. Detection of *B. burgdorferi* in human skin biopsies by a nested PCR with ospA-specific primers (lanes 1 and 2) or ospA-specific primers together with primers specific for the human ß$_2$-microglobulin-gene as an internal control (lanes 3 and 4). Each „lane" represents triplicate samples of one skin biopsy processed in parallel. Lanes 1 and 3: patient no. 2 with erythema migrans; lane 2 and 4: patient no. 3 with suspected erythema migrans; lanes without identification no.: TE-buffer used as a negative control; lane 5: 1 fg *B. burgdorferi* DNA used as a positive control; S: pBR328 BgI I + pBR328 Hinf I DNA-size standard. The triplicate sample of patient no. 2 (EM) resulted in an ospA-specific 392 bp band in both the PCR without (lane 1) and with (lane 3) internal control primers for ß$_2$-microglobulin . Samples of patient no. 3 (suspected EM) did not result in an ospA-specific band in both the PCR without (lane 2) and with (lane 4) internal control primers. In contrast, samples from both patients gave a ß$_2$-microglobulin-specific 677 bp band in the nested PCR containing ospA-specific primers and primers for the human ß$_2$-microglobulin-gene (lanes 2 and 4).

proteinase K digestion of samples has turned out to be sufficient to prepare human skin biopsies for PCR. We feel that internal control primers for the amplification of human DNA sequences (e.g. ß$_2$-microglobulin-gene-sequences) are necessary to assess the reliability of PCR results. For the interpretation of results the clinical diagnosis must be considered. Whenever possible, other methods (e.g. serological testing) should be used to provide supplementary evidence for a *B. burgdorferi* infection. Nevertheless one should bear in mind, that concordant results between PCR and serologic tests are not to be expected *a priori*, since different targets (direct demonstration of DNA versus indirect demonstration of an antibacterial antibody response) are detected by these tests.

REFERENCES

Güssow D, Rein R, Ginjaar I, Hochstenbach F, Seemann G, Kottmann A and Ploegh H L. The human ß$_2$-microglobulin gene - Primary structure and definition of the transcriptional unit. J Immunol 1987, 139: 3132-3138

Hinnebusch J and Barbour A G. Linear- and circular-plasmid copy numbers in *B. burgdorferi*. J Bacteriol 1992, 174: 5251-5257

Howe T R, Mayer L W and Barbour A G. A single recombinant plasmid expressing two major outer surface proteins of the Lyme disease spirochete. Science 1985, 227: 645-646

Schwan T G, Simpson W J, Rosa P A. Laboratory confirmation of Lyme disease. Can J Infect Dis 1991, 2: 64-69.

Steere A C. Lyme disease. N Engl J Med 1989, 321: 586-596

Wallich R , Moter SE, Kramer MD, Gern L, Hofmann H , Schaible U E, Simon M M. Untersu-
chungen zur genotypischen und phänotypischen Heterogenität von *Borrelia burgdorferi*, dem
Erreger der Lyme-Borreliose. in: Lyme-Borreliose: Fortschritte der Infektiologie. ed. Hassler,
D., München, MMV, Medizin-Verlag 1992. 167-190

THE APPLICATION OF PCR FINGERPRINTING TO THE EPIDEMIOLOGIC ANALYSIS OF BACTERIAL AND FUNGAL PATHOGENS

G. Schönian[1], Y. Gräser[1], O. Meusel[2], W. Meyer[3], P. Buchholz[1], W. Presber[1] and Th. G. Mitchell[3]

[1]Institut für Mikrobiologie und Hygiene, Universitätsklinikum Charité der Humboldt-Universität, Berlin; [2]Zentrum für Zahn-, Mund- und Kieferheilkunde, Universitätsklinikum Charité der Humboldt-Universität, Berlin; [3]Department of Microbiology, Duke University Medical Center, Durham, USA

INTRODUCTION

The increased incidence of nosocomial infections especially in immunocompromised patients has stimulated interest in developing more definite procedures for epidemiological studies of the etiological agents, such as pathogenic strains of different bacterial and fungal species. To determine the origins of infection, the routes of acquisition and transmission as well as the persistence of pathologic strains, precise and reproducible diagnostic methods are required. Recently, Williams et al (1990) and Welsh and McClelland (1990) described a PCR-based method for assessing DNA polymorphisms by amplifying genomic DNA with single primers of arbitrary nucleotide sequence. The use of the PCR with primers differing in length and nucleotide composition detected polymorphisms in the absence of specific sequence information in DNA from bacteria (Jayarao et al 1992), fungi (Crowhurst et al 1991), plants (Wilde et al 1992), animals (Welsh et al 1991) and humans (Williams et al 1990). Polymorphisms generated by this method have been termed arbitrarily primed-polymerase chain reaction (AP-PCR) fingerprinting (Welsh and McClelland 1990), random amplification of polymorphic DNA (RAPD) markers (Williams et al 1990) and DNA amplification fingerprinting (Jayarao et al 1992). The term PCR fingerprinting is used here because single primers , which were originally applied as hybridization probes in conventional DNA fingerprinting to detect minisatellite and microsatellite DNA such as the core sequence of phage M 13 (5'-GAGGGTGGCGGTTCT), and the simple repeat sequences $(GACA)_4$ and $(GATA)_4$ (Meyer et al.1991; Ali et al 1986) were used to amplify DNA sequences in the genome of different bacterial and yeast species. Besides, in our experiments primers were tested which were already used in other labs for AP-PCR (universal sequence of M 13; 5'-TTATGAAACGACGGCCAGT; Welsh et al 1991) or RAPD (10-mer oligonucleotide; 5'-TCACGATGCA; Williams et al).

PCR fingerprint procedure

PCR fingerprinting requires only small amounts of DNA which should be available from simple crude minipreparations. Bacterial DNA was extracted by a simple boiling method. 5 colonies were suspended in 100µl water or 50µl sodium hydroxide and boiled for 15 minutes. After neutralization, if alkali was used, and a brief centrifugation 2.5 µl of the crude extract were applied to the PCR. Yeast DNA was isolated using a standard protocol including phenol extraction which was originally designed for quantitative recovery of the DNA (Gruber, 1990). 25 ng of the purified yeast DNA was applied to the standard PCR assay. Genomic fingerprinting was achieved by setting up standard PCR reactions but with single arbitrary primers, somewhat higher magnesium and an altered thermocycling profile. Amplification reactions were performed in volumes of 50 µl containing template DNA as described above; 10 mM Tris/HCl, pH 8.3; 50 mM KCl; 1.5 mM $MgCl_2$; 3 mM Mg-acetate; 200 µM each of dATP, dCTP, dGTP and dTTP and 2.5 U Taq DNA Polymerase. The primers were added at a final concentration of 4 to 25 pmol. The annealing for all primers was performed at the melting temperature of the oligonucleotide used. The samples were amplified for 27 to 35 cycles as follows: 20 s at 95°C, 60 s at annealing temperature, 20 s at 72°C, followed by a final extension cycle of 6 min at 72°C. Amplified DNA fragments were separated by electrophoresis in a 1.2% agarose gel and detected by staining with ethidium bromide.

Distinct PCR patterns for all bacterial and fungal species tested were obtained employing the two M13 primers and the $(GACA)_4$ primer. The discriminating power of the core sequence of M13 and the $(GACA)_4$ primer seemed to be greater than that of the universal sequence of M13 since on average more fragments were amplified with PCR using these two primers. PCR fingerprinting could be successfully applied for strain typing of different bacteria (E.coli, Salmonella spp., Bacteroides spp. and other anaerobic bacteria, Serratia marcescens, Acinetobacter baumannii, Pseudomonas spp.) and of different Candida spp. (data not shown).

To evaluate the reproducibility of the PCR fingerprints DNA from different isolates was amplified in duplicate, and no differences were observed between products of duplicate samples. Furthermore, with the DNA extracted from the same Candida albicans isolate and amplified on three separate runs with primer M13, identical fingerprint patterns were generated (data not shown). Although, these DNA polymorphisms were reproducible, variations in the intensity of amplified bands were occasionally observed. Those differences may be attributed to slight variations in the temperature of individual wells in the thermocycler or to different amounts of template DNA in crude extracts.

In some cases bands could be detected in the control samples without DNA; however, these artifactual bands were clearly different from all the bands obtained with the various DNA samples. These artifacts probably resulted from the contamination of Taq polymerase with bacterial DNA (Böttger, 1990). A reduction of the number of PCR cycles to 27 led to the disappearance of these bands in the control sample (data not shown).

Epidemiological study of an *Acinetobacter baumannii* outbreak using PCR fingerprinting

PCR fingerprinting was applied to characterize 49 *Acinetobacter baumannii* isolates from an outbreak at our hospital, the Charité (Gräser 1993). A total of 13 patients of the anaesthesiology intensive care unit were infected or colonized with multiresistant strains of this pathogen. All patients were mechanically ventilated after neurosurgery. Seven of these patients had severe pneumonia; one developed urinary tract infection. Using the core sequence of M13 as a primer the PCR patterns of 45 isolates obtained from 12 patients that displayed similar antibiotic susceptibility were found to be identical. These findings suggest that 45 of the 49 isolates tested belonged to the same subtype of *Acinetobacter baumannii* (CH 1) and may have been epidemiologically related. The remaining 4 strains which were isolated from the same patient, however, exhibited a distinct amplification pattern and represented a different subtype (CH 2). Another five *Acinetobacter baumannii* isolates were obtained from 5 patients of different intensive care units from hospitals of Magdeburg and Oschersleben. The three isolates from the Hospital of Magdeburg (M) had the same amplification pattern which, however differed clearly from those of all other strains investigated. Individual PCR fingerprints were yielded for both strains from the hospital of Oschersleben (O 1 and O 2) as well as for the reference strain ATCC 19606 (Figure 1).

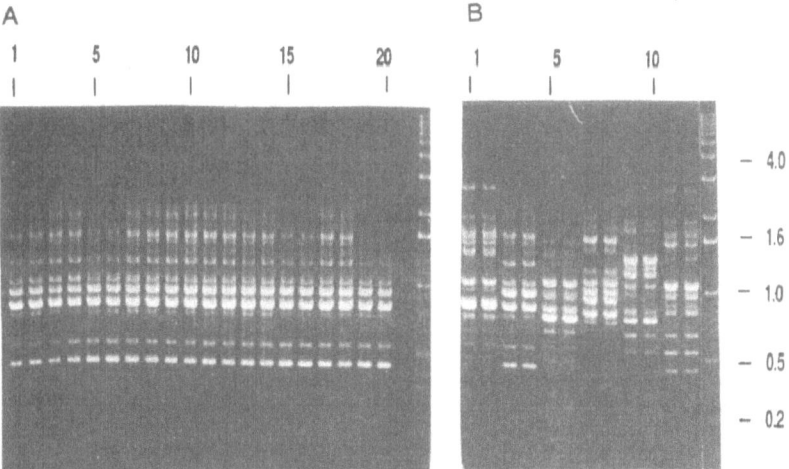

Figure1. M13 primed PCR fingerprints of 26 *Acinetobacter baumannii* isolates from different origins. a) 20 isolates of *Acinetobacter baumannii* obtained from 12 different patients belonging to the same genetic subtype CH 1. Lane 21 was a control sample without DNA, lane 22 molecular size markers in kb. b) PCR fingerprints of *Acinetobacter baumannii* strains belonging to different genetic subtypes. All samples were amplified in duplicate. Lanes 1 and 2, ATCC strain 19606; lanes, 3 and 4 subtype CH 1; lanes 5 and 6, subtype CH 2; lanes 7 and 8, subtype M (Magdeburg): lanes 9 and 10, subtype O 1 (Oschersleben); lanes 11 and 12, subtype O 2; lane 13 molecular size markers in kb.

A computer analysis of the genetic relationship using RFLPrint, version 1 software (SPARCstation IRC, Sun Microsystems, Inc., USA) revealed a homology of only 54% for the *Acinetobacter* subtypes isolated in the Charité hospital and only a low degree of genetic relatedness between the *Acinetobacter baumannii* strains of different geographical locations in this study.

Application of PCR fingerprinting to epidemiological analysis of *Serratia marcescens* strains

56 *Serratia marcescens* strains were isolated during outbreaks of nosocomial infections which occurred at different time in the neonatal intensive care unit of the Charité hospital. The patients suffered from septicemia or meningitis and the mortality was up to 75%. The strains of different outbreaks were previously typed by plasmid profile analysis (Buchholz 1990). 54 of these strains belonged to three types of plasmid profiles, the remaining two strains had no plasmids. Fingerprinting these strains with three different primers (core sequence of M13; universal sequence of M13; and the 10 mer oligonucleotide sequence) could distinguish four discrete types of amplification patterns (Figure 2). 77% of the strains could be assigned to the same genetic groups as compared with the plasmid profile analysis. 13 strains belonged, however, to one of the other three subtypes.

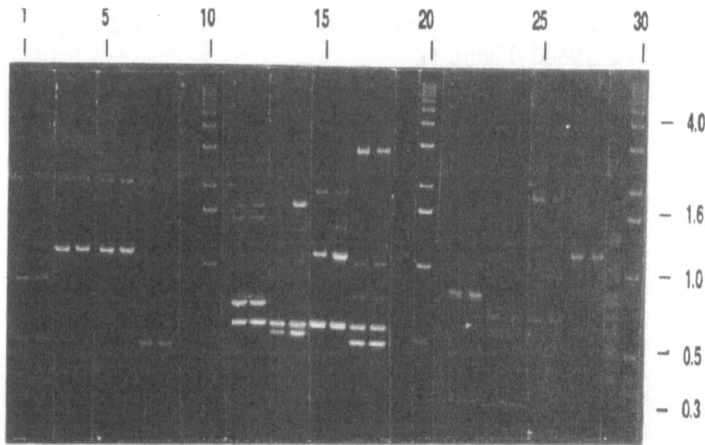

Figure 2. PCR fingerprints of *Serratia marcescens* strains belonging to four different genetic groups. Two *Serratia marcescens* strains of the same subtype were amplified with the universal sequence of M13 (lanes 1 to 8), with the core sequence of M13 (lanes 11 to 18) and with the 10mer oligonucleotide (lanes 21 to 28). Lanes 9, 19, and 29, are control samples without DNA, and lanes 10, 20, and 30 molecular size markers in kb. The cycle number was 27 if the M13 primers were used and 45 if the 10mer primer was applied.

PCR Fingerprinting of clinical *Candida albicans* strains

Species of *Candida* are a frequent source of hospital-acquired infections. At the Charité hospital approximately 80% of patients in the intensive care units are colonized by species of this yeast. About half of these patients develop systemic candidiasis, and the mortality in this group of patients is rather high. Many authors have postulated that this infection results from immunological impairment and is caused by endogenous strains of *Candida*, since isolates from different patients have not been related (Reagan et al 1990; Whelan et al 1990; Clemons et al 1991). However, recently reported outbreaks suggest that exogenous sources of *Candida* may be involved in the development of nosocomial candidiasis (Stevens et al 1990; Matthews and Burnie 1989; Schmid et al 1990).

The core sequence of phage M 13 and the simple repeat sequence $(GACA)_4$ were used to amplify DNA sequences in the genome of *Candida albicans* strains (Schönian et al 1993). The two primers generated different sets of PCR-fingerprint patterns, but the discriminating power of the primers was similar.

Since considerable polymorphisms were produced in *Candida* strains by PCR fingerprinting with single primers it was tried to compare isolates from patients on several intensive care units as well as from healthy dental clients. The PCR fingerprints of isolates from all patients from each ICU were highly similar (Figure 3). Isolates from two patients (lane 2 and 5) who were both located in ICU I and died following candidemia, had identical PCR fingerprints. Isolates from four patients (lane 13, 15, 17, and 18), in ICU V, all of whom were successfully treated during the same time frame, produced similar patterns. Isolates in the lanes 1, 7, and 16 which showed quite different PCR fingerprints were identified as *Candida glabrata*.

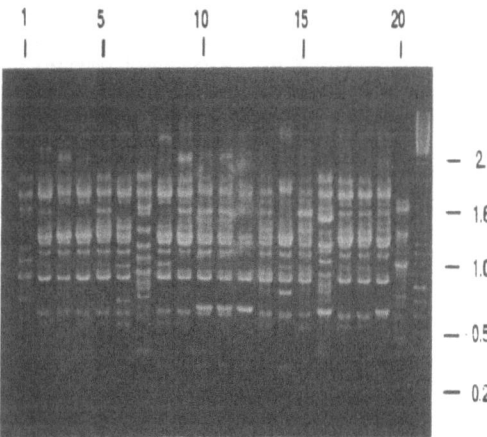

Figure 3. Comparison of M13-primed, PCR fingerprints among 19 isolates of *Candida albicans* from patients in different intensive care units (ICU). The isolates in lanes 1, 7, and 16 are *Candida glabrata*. All other isolates are *Candida albicans*. Lanes 1 to 6, isolates obtained from patients in ICU I (emergency room); lane 7, isolate obtained from a patient who was transferred from the ICU I to ICU N (neurology); lanes 8 to 12, isolates obtained from patients on ICU IVb (organ transplantation); lanes 13 to 19, isolates obtained from patients in ICU V (gastrointestinal surgery); lane 20, control sample without DNA; lane 21, molecular size markers in kb.

Multiple *Candida* isolates from different body sites of the same patient yielded either identical or different PCR fingerprints (data not shown). The same phenomenon was observed with healthy individuals (Schmid et a, 1990). The explanation for variations among isolates from the same person is not clear.

Compared with the isolates from the patients in the intensive care units, the *Candida albicans* isolates from healthy dental clients exhibited more genetic divergence (data not shown). Isolates from the dentures and the gums of the same client yielded identical PCR fingerprints, but the pattern of isolates from different subjects were clearly distinguishable.

CONCLUSION

PCR fingerprinting was successfully used for epidemiological studies of different pathogens causing nosocomial infections. Compared to other methods of DNA-based strain identification, PCR fingerprinting offers the advantages of simplicity and rapidity. Using a simple method of DNA extraction from bacteria it was possible to differentiate about 40 strains in two days. Hence, PCR-fingerprinting may prove very useful for surveying large numbers of isolates not only for epidemiological but also for taxonomic and population studies.

ACKNOWLEDGEMENT

We thank E. Halle and H.J. Tietz for supplying *Acinetobacter baumannii, Serratia marcescens* and *Candida albicans* strains as well as for helpful discussions. This research was supported by Fonds der Chemischen Industrie, Frankfurt/Main, by the Deutsche Forschungs-gemeinschaft (Bu 815/1-1) and by a grant # AI28836 from the National Institutes of Health, U.S. Public Health Service.

REFERENCES

Ali S, Müller CR and Epplen JT. DNA fingerprinting by oligonucleotide probes specific for simple repeats. Hum. Genet. 1986, 74: 239-243

Böttger EC. Frequent contamination of Taq polymerase with DNA. Clin. Chem. 1990, 36: 1258-1259

Buchholz P, Prager R, Tschäpe H, Halle E, Grauel EL and Schmidt G. *Serratia marcescens* als Ursache eines Ausbruchs von septikämischen Infektionen. Z. Klin. Med. 1990, 45: 2035-2037

Clemons KV, Shankland GS, Richardson MD and Stevens DA. Epidemiologic study by DNA typing of a *Candida albicans* outbreak in heroin addicts. J. Clin. Microbiol. 1991, 29: 205-207

Crowhurst RN, Hawthorne BT, Rikkerink EHA and Templeton MD. Differentiation of *Fusarium solani f. sp. cucurbitae* races 1 and 2 by random amplification of polymorphic DNA. Curr. Genet. 1991, 20: 391-396

Gräser Y, Klare I, Halle E, Gantenberg R, Buchholz P, Presber W and Schönian G. Epidemiological study of an *Acinetobacter baumannii* outbreak using PCR fingerprinting.J ClinMicrobiol 1993, 31: 2417-2420

Gruber F. Homologe und heterologe Transformation von *Trichoderma reesei* mit den Ornithin-5'-Decarboxylase-Genen als Selektionsmarker. Ph.D. Thesis. TU Wien

Jayarao BM, Bassam BJ, Caetano-Anolles G, Gresshoff PM and Oliver SP. Subtyping of *Streptococcus uberis* by DNA amplification fingerprinting. J. Clin. Microbiol. 1992, 30: 1347-1350

Matthews R and Burnie J. Assessment of DNA fingerprinting for rapid identification of outbreaks of systemic candidiasis. Br. Med. J. 1989, 298: 354-357

Meyer W, Koch A, Niemann C, Beyermann B, Epplen JT and Börner T. Differentiation of species and strains among filamentous fungi by DNA fingerprinting. Curr. Genet. 1991, 19: 239-242

Reagan DR, Pfaller MA, Hollis RJ and Wenzel RP. Characterization of the sequence of colonization and nosocomial candidemia using DNA fingerprinting and a DNA probe. J. Clin.Microbiol. 1990, 28: 2733-2738

Schmid J, Voss E and Soll DR. Computer-assisted methods for assessing strain relatedness in *Candida albicans* by fingerprinting with the moderately repetitive sequence Ca3. J. Clin. Microbiol. 1990, 28: 1236-1243

Schönian G, Meusel O, Tietz HJ, Meyer W, Gräser Y, Tausch I, Presber W and Mitchell TG. Identification of clinical *Candida albicans* by DNA fingerprinting with the polymerase chain reaction. Mycoses 1993, 36: 171-179

Stevens DA, Odds FC and Scherer S. Application of DNA typing methods to *Candida albicans* epidemiology and correlations with phenotype. Rev. Infect. Dis. 1990, 12: 258-266

Welsh J and McClelland M. Fingerprinting genomes using PCR with arbitrary primers. Nucleic Acids Res. 1990, 18: 7213-7218

Welsh J, Petersen C and McClelland M. Polymorphisms generated by arbitrarily primed PCR in the mouse: application to strain identification and genetic mapping. Nucleic Acids Res. 1991, 19: 303-306

Whelan WL, Kirsch DR, Kwon-Chung KJ, Wahl SM and Smith PD. *Candida albicans* in patients with the acquired immunodeficiency syndrome: absence of a novel or hypervirulent strain. J.Infect. Dis. 1990, 162: 513-518

Wilde J, Waugh R and Powell WA. Genetic fingerprinting of *Theobrama* clones using randomly amplified polymorphic DNA. Theor. Appl. Genet. 1992, 83: 871-877

Williams JGK, Kubelik AR, Livak KJ, Rafalski JA and Tingey SV. DNA polymorphisms amplified by arbitrary primers are useful as genetic markers. Nucleic Acids Res. 1990, 18: 6531-6535

CONTRIBUTORS

Altwegg, Martin
Institute of Medical Microbiology, University of Zürich, CH-8028 Zürich, Switzerland.

Anson, J G
Amersham International PLC, Cardiff Laboratorie, Forest Farm, Whitchurch, Cardiff CF4 7YT, UK.

Aouizenate, A
Laboratory LCL, 37 Street Boulevard, Paris 14^{éme}, France.

Bacher, Michael
Institute of Immunology, Philipps University, Robert-Koch-Straße 17, 35037 Marburg, Germany.

Banatvala, J E
Department of Virology, United Medical and Dental Schools of Guy's & St Thomas'Hospitals, (St Thomas'Campus), Lambeth Palace Road, London, UK.

Barany, Francis
Department of Microbiology, Cornell University Medical College, New York, USA.

Batt, Carl A
Department of Food Science, Cornell University, Ithaca, New York, USA.

Bazubagira, Anatholie
Department of Paediatrics, Centre Hospitalier de Kigali, Rwanda.

Becker, K
Institute of Medical Microbiology, Medical Academy Magdeburg, Leipziger Straße 44, 39120 Magdeburg, Germany.

Böhtig, B
Robert-Koch-Institut des Bundesgesundheitsamtes, Corrensplatz 1, 14195 Berlin, Germany.

Brade, V
Abteilung Mikrobiologie im Zentrum für Hygiene der Universitätskliniken Frankfurt, Germany.

Brandt, B
Institute of Clinical Chemistry, WWU Münster, Albert-Schweitzer-Straße 3, 48149 Münster, Germany.

Brede, H D
Chemotherapeutisches Forschungsinstitut Georg-Speyer-Haus, Paul-Ehrlich-Straße 42-44, Frankfurt am Main, Germany.

Brodt, R
Zentrum Innere Medizin, Medizinische Universitätskliniken Frankfurt, Germany.

Buchholz, Petra
Institut für Mikrobiologie und Hygiene, Universitätsklinikum Charité der Humboldt-Universität, Clara-Zetkin-Straße 96, PF 140, Berlin, Germany.

Champenois, T
Laboratory LCL, 37 Street Boulevard, Paris 14éme, France.

Cross, L
Amersham International PLC, Cardiff Laboratories, Forest Farm, Whitchurch, Cardiff CF4 7YT, UK.

Czajka, John
Department of Food Science, Cornell University, Ithaca, New York, USA.

Dabis, Francois
INSERM U330, University of Bordeaux II, France.

Dahme, Miriam
Institute of Medical Microbiology, University of Greifswald, Martin-Luther-Straße 6, 17489 Greifswald, Germany.

Decker, D J
Scripps Research Institute, 10666 North Torrey Pines Road, La Jolla, CA 92037, USA.

Diedrich, S
Robert-Koch-Institut des Bundesgesundheitsamtes, Corrensplatz 1, 14195 Berlin, Germany.

Ellis, J
Central Public Health Laboratory, 61 Colindale Avenue, London NW9 5 HT, UK.

Flehmig, Bertram
Department of Medical Virology and Epidemiology of Virus Diseases, Hygiene-Institute, University of Tübingen, Silcher Straße 7, 72076 Tübingen, Germany.

Flunker, Gisela
Klinik und Poliklinik für Kindermedizin, Ernst Moritz-Arndt Universität, 17489 Greifswald, Germany.

Gemsa, Diethard
Institute of Immunology, Philipps University, Robert-Koch-Straße 17, 35037 Marburg, Germany.

Gerth, Hans-Joachim
Department of Medical Virology and Epidemiology of Viral Diseases, Hygiene-Institute, University of Tübingen, Silcherstr. 7, 72076 Tübingen, Germany.

Gingeras, T
Baxter Diagnostics Incorporation, Life Sciences Research Laboratory, 4245 Sorrento Valley Boulevard, San Diego, CA 92121, USA.

Gräser, Yvonne
Institut für Mikrobiologie und Hygiene, Universitätsklinikum Charité der Humboldt-Universität, Clara-Zetkin-Straße 96, PF 140, Berlin, Germany.

Griwatz, C
Institute of Clinical Chemistry, WWU Münster, Albert-Schweitzer-Straße 3, 48149 Münster, Germany.

Hamprecht, Klaus
Department of Medical Virology and Epidemiology of Viral Diseases, Hygiene-Institute, University of Tübingen, Silcherstr. 7, 72076 Tübingen, Germany.

Handschack, A
Institute of Medical Microbiology, Medical Academy Magdeburg, Leipziger Straße 44, 39120 Magdeburg, Germany.

Harms, F
Institute of Clinical Chemistry, WWU Münster, Albert-Schweitzer-Straße 3, 48149 Münster, Germany.

Hartl, M
Boehringer Mannheim GmbH, Nonnenwald 2, 82377 Penzberg, Germany.

Helm, E B
Zentrum Innere Medizin, Medizinische Universitätskliniken Frankfurt, Germany.

Hitimana, Deo-Gratias
Department of Paediatrics, Centre Hospitalier de Kigali, Rwanda.

Hofmann, Peter
Institute of Immunology, Philipps University, Robert-Koch-Straße 17, 35037 Marburg, Germany.

Kaletta, C
Boehringer Mannheim GmbH, Nonnenwald 2, 82377 Penzberg, Germany.

Kämmerer, Ute
Medizinische Klinik II, University of Erlangen, Östliche Stadtmauerstraße 29, 91054 Erlangen, Germany.

Karita, Etienne
National AIDS Control Programme, Aids Reference Laboratory, Kigali, Rwanda.

Kessler, H H
Institute of Hygiene, KF-University of Graz, Universitaetsplatz 4, 8010 Graz, Austria.

Kievits, T
Organon Teknika, Boseind 15, 5281 RM Boxtel, The Netherlands.

Kirch, P
Boehringer Mannheim GmbH, Nonnenwald 2, 82377 Penzberg, Germany.

Kleiber, J
Boehringer Mannheim GmbH, Nonnenwald 2, 82377 Penzberg, Germany.

Kleinhappl, B
Institute of Hygiene, KF-University of Graz, Universitaetsplatz 4, 8010 Graz, Austria.

Klinman, N R
Scripps Research Institute, 10666 North Torrey Pines Road, La Jolla, CA 92037, USA.

Koch, G
Institute of Medical Microbiology, Medical Academy Magdeburg, Leipziger Straße 44, 39120 Magdeburg, Germany.

Korge, Bernhard P
Klinik für Dermatologie der Universität Köln, Joseph-Stelzmannstr. 9, 50931 Köln, Germany.

Korn, K
Institut für Klinische und Molekulare Virologie, University of Nürnberg-Erlangen, 91054 Erlangen, Germany.

Koutz, P
Baxter Diagnostics Incorporation, Life Sciences Research Laboratory, 4245 Sorrento Valley Boulevard, San Diego, CA 92121, USA.

Kramer, Michael D
Institute for Immunology, University of Heidelberg, Heidelberg, Germany.

Kunkel, B
Medizinische Klinik II, University of Erlangen, Östliche Stadtmauerstr. 29, 91054 Erlangen, Germany.

Lens, Peter F
Organon Teknika, Boseind 15, 5281 RM Boxtel, The Netherlands.

Lepage, Philippe
Department of Paediatrics, Centre Hospitalier de Kigali, Rwanda.

Linton, P-J
Scripps Research Institute, 10666 North Torrey Pines Road, La Jolla, CA 92037, USA.

Löning, Thomas
Department of Gynaecological Pathology and Electron Microscopy, Clinics of Obstetrics and Gynaecology, University of Hamburg, Germany.

López-Pila, J M
Institut für Wasser-, Boden- und Lufthygiene des Bundesgesundheitsamtes, Corrensplatz 1, 14195 Berlin, Germany.

Lorenz, G
Institute of Pathology, University of Greifswald, Fleischmannstraße, 17489 Greifswald, Germany.

Lucotte, G
Laboratory of Molecular Anthropology, CHU of Cochin Port-Royal, 24 Street of Fauburg St Jacques, Paris 14éme, France.

Majewski, C
Boehringer Mannheim GmbH, Nonnenwald 2, 82377 Penzberg, Germany.

Marchand, J
CIS BIO International, BP 32, Gif-sur-Yvette, 91192, France.

Marth, E
Institute of Hygiene, KF-University of Graz, Universitaetsplatz 4, 8010 Graz, Austria.

Martinetti Lucchini, Gladys
Institute of Medical Microbiology, University of Zürich, 8028 Zürich, Switzerland.

Meusel, Olaf
Zentrum für Zahn-, Mund- und Kieferheilkunde, Universitätsklinikum Charité der Humboldt-Universität, Schumannstraße 20/21, PF 140, Berlin, Germany.

Meyer, Wieland
Department of Microbiology, Duke University Medical Center, Durham, NC, 27710, USA.

Mischke, Dietmar
Institut für Experimentelle Onkologie und Transplantationsmedizin, Universitätsklinikum Rudolf Virchow, Freie Universität Berlin, Spandauer Damm 130, 14050 Berlin, Germany.

Mitchell, Thomas G
Department of Microbiology, Duke University Medical Center, Durham, NC, 27710, USA.

Moter, Sabine E
Institute for Immunology, University of Heidelberg, Heidelberg, Germany.

Msellati, Phillippe
INSERM U330, University of Bordeaux II, France.

Muir, Peter
Department of Virology, United Medical and Dental Schools of Guy's & St Thomas'Hospitals (St Thomas'Campus), Lambeth Palace Road, London, UK.

Müller, W A
Institute of Medical Microbiology, Medical Academy Magdeburg, Leipziger Straße 44, 39120 Magdeburg, Germany.

Mura, C
Laboratory INSERM Y 120, Robert Dabré Public Hospital, 48 Boulevard Serurier, Paris, 19^{éme}, France.

Myerson, David
University of Washington, Fred Hutchinson Cancer Research Center, Seattle, USA.

Nicholson, F
Department of Virology, United Medical and Dental Schools of Guy's & St Thomas'Hospitals, (St Thomas'Campus), Lambeth Palace Road, London, UK.

Nsemgumuremyi, Francois
National AIDS Control Programme, Aids Reference Laboratory, Kigali, Rwanda.

O'Leary, John
Nuffield Department of Pathology and Bacteriology, University of Oxford, UK

Orchert, A
King's College School of Medicine and Dentistry, London, UK.

Peters, Angelika
Institute of Medical Microbiology, University of Greifswald, Martin-Luther-Straße 6, 17489 Greifswald, Germany.

Philipp, E
Institute of Medical Microbiology, Medical Academy Magdeburg, Leipziger Straße 44, 39120 Magdeburg, Germany.

Pierer, K
Institute of Hygiene, KF-University of Graz, Universitaetsplatz 4, 8010 Graz, Austria.

Potts, C
Amersham International PLC, Cardiff Laboratories, Forest Farm, Whitchurch, Cardiff CF4 7YT, UK.

Presber, Wolfgang
Institut für Mikrobiologie und Hygiene, Universitätsklinikum Charité der Humboldt-Universität, Clara-Zetkin-Straße 96, PF 140, Berlin, Germany.

Pschaid, A
Institute of Hygiene, KF-University of Graz, Universitaetsplatz 4, 8010 Graz, Austria.

Reddemann, Hans
Institute of Medical Microbiology, University of Greifswald, Martin-Luther-Straße 6, 17489 Greifswald, Germany.

Riethdorf, L
Institute of Pathology, University of Greifswald, Fleischmannstraße, 17489 Greifswald, Germany.

Riethdorf, S
Institute of Microbiology, University of Greifswald, 17489 Greifswald, Germany.

Rolfs, Arndt
Arbeitsgruppe für Molekulare Neurobiologie, Institut für Neuropsychopharmakologie der FU Berlin, Ulmenallee 30, 14050 Berlin, Germany.

Rübsamen-Waigmann, Helga
Chemotherapeutisches Forschungsinstitut Georg-Speyer-Haus, Paul-Ehrlich-Straße 42-44, Frankfurt am Main, Germany.

Schneider, C
Abteilung Mikrobiologie im Zentrum für Hygiene der Universitätskliniken Frankfurt, Germany.

Schönian, Gabriele
Institut für Mikrobiologie und Hygiene, Universitätsklinikum Charité der Humboldt-Universität, Clara-Zetkin-Straße 96, PF 140, Berlin, Germany.

Schreier, E
Robert-Koch-Institut des Bundesgesundheitsamtes, Corrensplatz 1, 14195 Berlin, Germany.

Schweiger, B
Robert-Koch-Institut des Bundesgesundheitsamtes, Corrensplatz 1, 14195 Berlin, Germany.

Seidel, Werner
Institute of Medical Microbiology, University of Greifswald, Martin-Luther-Straße 6, 17489 Greifswald, Germany.

Simon, Markus M
Max-Planck-Institute for Immunobiology, Freiburg, Germany.

Simonon, Arlette
National AIDS Control Programme, Aids Reference Laboratory, Kigali, Rwanda.

Slomka, M J
Central Public Health Laboratory, 61 Colindale Avenue, London NW9 5 HT, UK.

Stegner, H E
Department of Gynaecological Pathology and Electron Microscopy, Clinics of Obstetrics and Gynaecology, University of Hamburg, Germany.

Stillman, C
Baxter Diagnostics Incorporation, Life Sciences Research Laboratory, 4245 Sorrento Valley Boulevard, San Diego, CA 92121, USA.

Stünzner, D
Institute of Hygiene, KF-University of Graz, Universitaetsplatz 4, 8010 Graz, Austria.

Teo, C G
Central Public Health Laboratory, 61 Colindale Avenue, London NW9 5 HT, UK.

Thums, U
Chemotherapeutisches Forschungsinstitut Georg-Speyer-Haus, Paul-Ehrlich-Straße 42-44, Frankfurt am Main, Germany.

van de Perre, Philippe
National AIDS Control Programm, Aids Reference Laboratory, Kigali, Rwanda.

van Gemen, R
Organon Teknika, Boseind 15, 5281 RM Boxtel, The Netherlands.

van Goethem, Christiaan
Department of Paediatrics, Centre Hospitalier de Kigali, Rwanda.

Vogt, U
Institute of Clinical Chemistry, WWU Münster, Albert-Schweitzer-Straße 3, 48149 Münster, Germany.

Wallich, Reine
German Cancer Research Center, Heidelberg, Germany.

Wang, Chwan-Heng
Department of Medical Virology and Epidemiology of Virus Diseases, Hygiene-Institute, University of Tübingen, Silcher Straße 7, 72076 Tübingen, Germany.

Wanner, Reinhard
Institut für Experimentelle Onkologie und Transplantationsmedizin, Universitätsklinikum Rudolf Virchow, Freie Universität Berlin, Spandauer Damm 130, 14050 Berlin, Germany.

Weber-Rolfs, Ines
Arbeitsgruppe für Molekulare Neurobiologie, Institut für Neuropsychopharmakologie der FU Berlin, Ulmenallee 30, 14050 Berlin, Germany.

Wiedmann, Martin
Department of Food Science, Cornell University, Ithaca, New York, USA.

Wiersbitzky, Siegfried
Institute of Medical Microbiology, University of Greifswald, Martin-Luther-Straße 6, 17489 Greifswald, Germany.

Wilson, Wendy
Department of Plant Pathology, Cornell University, Geneva, New York, USA.

Zänker, K S
Institute of Immunology, University Witten-Herdecke, Stockumer Straße 10, 58453 Witten, Germany.

SUPPLIERS OF SPECIALIST ITEMS

Advanced Magnetics, Inc., 61 Mooney Street, Cambridge, MA 02138, USA

Aldrich Chemical Co., 940 West Saint Pual Avenue, Milwaukee, WI 53223, USA; The Old Brickyard, New Road, Gillinghamn, Dorset, SP8 4JL, UK

Ambion Inc., 2130 Woodward St.#200, Austin, Texas 78744-1832, USA

Amersham International PLC, Life Sciences Business, 1 Amersham Place, Little Chalfont, Bucks, HP79NA, UK; Amersham Buchler GmbH &Co.KG, Gieselweg 1, 38110 Braunschweig, Germany

Amicon Div. W.R.Grace & Co.-Conn, 72 Cherry Hill Drive, Beverly, MA, 01915, USA; Upper Mill, Stonehause, Gloucester, GL 10 2BJ, UK

Applied Biosystems, Inc., 850 Lincoln Center Drive, Foster City, CA 94404, USA. Biotech Instruments Limited, Unit A, Caxton Hill Extension Road, Caxton Hill, Hertford, SG13 7LS; UK

Bachofer GmbH, P.O.Box 7058, D-7410 Reutlingen, Germany

Beckman Instruments, Inc., 2500 Harbor Blvd. Fullerton, CA 92634, USA; Frankfurter Ring 115, 80807 München, Germany; 2500 Harbour Boulevard, PO Box 3100, Fullerton, CA 92634, USA

Becton Dickinson, Immunocytometry Systems, 2350 Qume Dr., San Jose, CA 95131, USA; Tullastr. 8-12, 69126 Heidelberg, Germany

Biochrom KG, Leonorenstr. 2-6, 12247 Berlin, Germany

Bio-Med, Gesellschaft für Biotechnologie, Schloß Ditfurth, 97531 Theres, Germany

Biometra, PO Box 157, Maidstone, Kent, ME14 2AT, UK

Bio-Rad Laboratories/Chemical Div., 3300 Regatta Boulevard, Richmond, CA 94804, USA

Boehringer Mannheim Biochemicals, P.O.Box 50414, Indianapolis, IN 46250, USA; Sandhofer Str. 116, 68305 Mannheim, Germany

Cangene Corporation, 3403 American Drive, Mississauga, Ontario L4V1T4, Canada

Cetus, see Perkin Elmer Cetus.

Clontech Laboratories, Inc., 4030 Fabian Way, Palo Alto, CA 94303, USA

Corbett Research, distributed in Europe by Labortechnik Fröbel, Hannoversche Straße 27a, CA 91311, USA

Coy Coporation, 22 Metty Drive, Ann Arbor, Michigan 48103, USA

Diagen GmbH, Niederheider Str. 3, 40589 Düsseldorf; Qiagen Inc., 9259 Eton Ave, Chatsworth,

Drummond Scientific Company, 500 Parkway, Box 700, Broomall, PA 19008, USA

DuPont de Nemours GmbH, Du-Pont-Str. 1, 61352 Bad Homburg, Germany; Biotechnology Systems Division, BRML, G-50986 Wilmington, DE 19898, USA

Dunn Labortechnik GmbH, Postfach 1104, D-5464 Asbach, Germany

Dynal AS, P.O.Box 158 Skoyen, N-0212 Oslo, Norway; 475 Northern Blvd., Great Neck, NY 11021, USA

Epicentre Technologies, 1202 Ann Street, Madison, Wi 53713, USA

Eppendorf-Netheler-Hinz GmbH, Barkhausenweg 1, 22339 Hamburg, Germany; 45635 Northport Loop East, Fremont, CA 94538, USA

Ericomp, 6044 Cornerstone Court West, Suite E, San Diego, California 92121, USA

Fluka Chemical Corp., 980 S. Second Street, Ronkonkoma, NY 11779-7238, USA

FMC BioProducts, 5 Maple Street, Rockland, ME 04841, USA; FMC BioProducts Europe, Risingevej 1, DK-2665 Vallensbaek Strand, Denmark

Fröbel Labortechnik, Hannoversche Straße 27a, 10115 Berlin, Germany

Gene-Trak Systems, Framingham, Massachusetts, USA

GIBCO BRL Life Technologies, Inc., Industrial Bioproducts, P.O.Box 6009, Gaithersburg, MD 20877, USA; Life Technologie Ltd., European Division, P.O.Box 35, Trident House, Renfrew Road, Paisley, PA3 4EF, Scotland

Gilson France SA, Box 27, 3000 West Beltline Hwy., Midleton, WI 53562, USA; 72 rue Gambetta, BP 45, F-95400

Glen Research Corporation, 44901 FalconPlace, Sterling, VA 22170, USA

Greiner GmbH, Postfach 1162, D-7743 Frickenhausen, Germany

Grant Instruments (Cambridge) Ltd., Barrington, Cambridge CB2 5QZ, UK

Heraeus Instruments, Inc., 111-A Corporate Blvd., S. Plainfield, NJ07080, USA

Hoefer Scientific Instruments, 654 Minnesota Street, Box 77387, San Francisco, California 94107-0387, USA; Unit 12, Croft Road Workshops, Croft Road, Newcastle under Lyme, ST5 0TH, UK

Hoffmann-La Roche AG, Diagnostica, Emil-Barell-Str. 1, 79639 Grenzach-Wyhlen, Germany

Hybaid National Labnet, P.O.Box 841, Woodbridge, NJ 07095, USA; 11-113 Waldegrave Road, Teddington, Middlesex, TW11 8LL, UK

ICN Biochemicals, PO Box 28050, Cleveland, DH 44128, USA; Eagle House, Peregrine Business Park, Gomm Road, High Wyacombe, Bucks, HP13 7DL, UK

Idaho Technology, USA, phone: 001-208-5246354, USA

Invitrogen Corporation, 11588 Sorrento Valley Road #20, San Diego, CA 92121, USA

Knauer, Geräte GmbH & Co KG, Heuchelheimer Straße 9, 61348 Bad Homburg v.d.H., Germany

Lark Sequencing Technologies Inc., 9545 Katy Freeway, Suite 200, Housten TX 77024-9870, USA

E. Merck, Reagents Division, Frankfurter STr. 250, P.O.Box 4119, D-6100 Darmstadt, Germany

MilliGen/Biosearch, Division of Millipore, 1986 Middlesex Turnpike, Burlington, MA 01803, USA

Millipore Ltd., The Boulevard, Blackmoor Lane, Watford, Hertfordshire, WD12RA, UK; Hauptstr. 87, 65760 Eschborn, Germany

MJ Research Inc., 24 Bridge Street, Watertown, Massachusetts 02172, USA

National Biosciences, 3650 Annapolis Lane, Plymouth, MN 55447, USA

New England Biolabs, 32 Tozer Road, Beverly, MA 01915-9990, USA; Postfach 2750, 6231 Schwalbach/Taunus, Germany; CP Laboaratories, PO Box 22, Bishop's Stortford, Herts, CM23 3DX, UK

Novagen, 565 Science Dr., Madison, Wi 53711, USA

A/S Nunc, P.O.Box 280, Kamstrup, DK-4000 Roskilde, Denmark; Hagenauer Str. 21a, 65203 Wiesbaden, Germany

Organon Teknika, Boseind 15, 5281 RM Boxtel, The Netherlands

Peninsula Laboratories, Inc., 611 Taylor Way, Belmont, CA 94002, USA; Neckarstaden 10, 69117 Heidelberg, Germany

Perkin-Elmer Cetus Instruments, 761 Main Avenue, Norwalk, CT 06859-0251, USA; P.O.Box 101164, D-7770 Überlingen, Germany

Pharmacia LKB Biotechnology AB, Bjorkgatan 30, Uppsala, Sweden; 800 Centennial Avenue, Piscataway, NJ 08855-1327, USA; Munzinger Str. 9, 79111 Freiburg, Germany

Polaroid Corp., Technical Imaging Products, 575 Technology Square, Cambridge, MA 02139, USA; Ashley Road, St. Albans, Herts, AL1 5PR, UK

Polygen GmbH, Karlstr. 10, 63225 Langen, Germany

Promega Corp., 2800 Woods Hollow Road, Madison, WI 53711-5399, USA

Savant Instruments Inc., 110-103 Bi-County-Blvd., Framingdale, NY 11735, USA

Schleicher & Schüll, Inc., 10 Optical Avenue, Keene, NH 03431, USA; P.O.Box 4, 37586 Dassel, Germany

Serva Feinbiochemica GmbH & Co., Carl-Benz-Str. 7, 69115 Heidelberg, Germany; 50 A & S Drive, Paramus, NJ 07652, USA

Sigma Chemical Co., P.O.Box 14508, St. Louis, MO 63178, USA; Francy Road, Poole, Dorset BH17 7NH, UK

Stratagene, 11099 N. Torrey Pines Rd., La Jolla, CA 92037, USA; P.O.Box 105466, 69121 Heidelberg, Germany

Synthetic Genetics, 3347 Industrial Court, San Diego, CA 92121, USA

Techne Incorporated, 3700 Brunswick Pike, Princeton, NJ 08540, USA

Tri-Continent Scientific Inc., 12555 Loma Rica Drive, Grass Valley, CA 95945, USA

Tropix Incorporated, 47 Wiggins Avenue, Bedford, Massachusetts 01730, USA

United States Biochemical Corp., P.O.Box 22400, Cleveland, OH 44122, USA; P.O.Box 2561, 61348 Bad Homburg, Germany

Whatman BioSystems Inc., 22 Bridewell Place, Clifton, NJ 07014, USA; Springfield Mill, Maidstone, Kent, ME14 2LE, UK

INDEX